CIRCULAR
EVIDENCE

Pat Delgado & Colin Andrews

CIRCULAR EVIDENCE

A detailed investigation of the flattened swirled crops phenomenon

BLOOMSBURY

First published 1989
Copyright © 1989 by Pat Delgado and Colin Andrews

Bloomsbury Publishing Ltd, 2 Soho Square, London W1V 5DE

British Library Cataloguing in Publication Data
Delgado, Pat
  Circular evidence.
  1. Unidentified flying objects
  I. Title II. Andrews, Colin
  001.9'42

ISBN 0−7475−0357−5

Designed by Malcolm Smythe

Studio work by Keith Shannon

Illustrated by Rob Shone

Typeset in Bodoni and Ehrhardt by Bookworm

Printed in Great Britain by Butler & Tanner Ltd, Frome, Somerset

In the summer of 1981 Pat Delgado brought to the attention of the national press the existence of some mysterious circular depressions in the fields at Cheesefoot Head, Hampshire. Since then, the sitings of similar phenomena in southern Britain – many of them striking in their symmetry and beauty – have transformed initial curiosity into a full-blown investigation. One look at the photographs will suggest why the circles raised so many questions – such as, how on earth did they get there?

The authors, Pat Delgado and Colin Andrews, began to compile case studies, conducting interviews with farmers and other witnesses and photographing the formations from hillsides, from the air and in close-up. With every aerial viewing and location visit, every measurement and close examination, more curious details seemed to emerge – details of site groupings and of the swirled and flattened circle floor formations, and the fact that scarcely any of the plants involved were damaged. Reports of further anomalous occurrences, mysterious coincidences possibly connected with the circles and related observations from other countries are also included.

While theories have naturally abounded since the circles came to light – theories that the team have considered and tested – the book does not set out to prove anything beyond the existence of a persistent and compelling enigma in our midst.

To Norah and Wendy,
for your patience and involvement,
and Jan,
for typing and assistance in production of this book.
With special thanks to Busty Taylor
for many superb photos,
and his wife Kathy;
also to Don Tuersly,
for his interest and observations.

ACKNOWLEDGEMENTS

We wish to thank all the farmers who were interested in this subject and kindly allowed us to enter their property; Mr Hall of Corhampton, Hampshire, Mr Botting of Longwood Estate, Hampshire, Mr Grainger of Chilcomb, Hampshire, Mr Stainer of South Wonston, Hampshire, Mr Brown of Headbourne Worthy, Hampshire, Mr Gibbons of Goodworth, Clatford, Hampshire, Mr Tippets of Upton Scudamore, Wiltshire, Mr Scull of Bratton, Wiltshire, Mr Cooper of Bratton, Wiltshire, Mr Gale of Bratton, Wiltshire, Mr Bevan of Old Alresford, Hampshire, Mr Matthews of Childrey, Oxfordshire, Mr Watson of Stoughton, Leicestershire, Mrs R. and J. Snook of Lavington, Wiltshire, Mr Partridge of Yatesbury, Wiltshire, Mr Flambert of Kimpton, Hampshire, Mr Horton of Beckhampton, Wiltshire, Mr Hughes of Silbury, Wiltshire, Mr Wills of Litchfield, Hampshire and Mr Sheppard of Bratton, Wiltshire. We would also like to thank all those people who have permitted us to enter their properties during our research, but who are not mentioned here.

Grateful acknowledgement is made to the following for permission to reprint their letters: B. Squires of Sydney, Bill Himmerman of Sarnia, Ontario, Colleen Kessler of Saskatchewan, Hugh Cochrane of Toronto, William A. Laux of Fauquier, J. E. Butler of Ottawa, Colin E. Chaplin of Belleville, Dennis J. Laughton of Calgary, Alberta, Alfred H. Beck of Pembroke, Ontario.

The authors are grateful to the following for permission to reproduce photographs: Nigel Taylor, Tim Martin, Ian Stevens, Barbara Hall, James Mathews, the *Andover Advertiser*, the British Ministry of Defence, Busty Taylor. All other photographs the authors' own.

The known is finite, the unknown infinite;
intellectually we stand on an islet in the midst of
an illimitable ocean of inexplicability. Our business
in every generation is to reclaim a little more land.

T. H. HUXLEY, 1825 — 95

CONTENTS

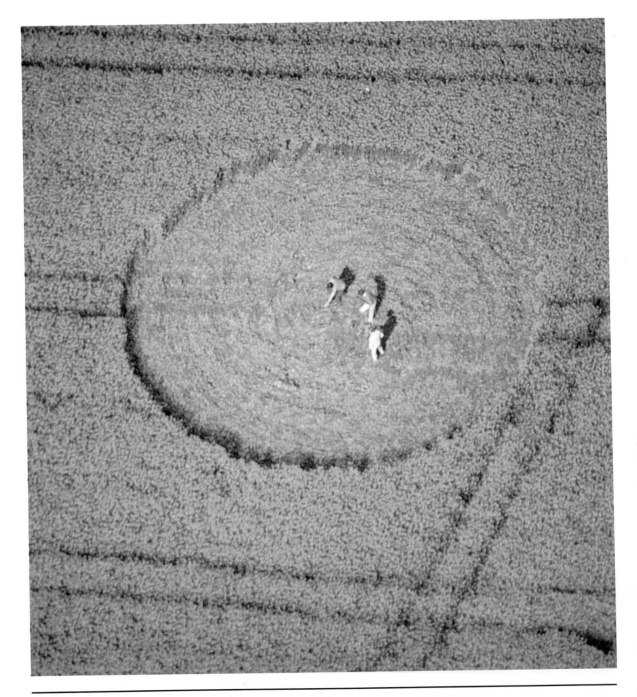

INTRODUCTION

The world is full of mysteries and will be into the foreseeable future. Various mysteries fascinate different people, but crop circles, the subject of this book, have fired the imaginations of a large cross-section of the population in many countries. Because of the great interest they arouse, our small group of investigators has decided to publish the detailed information we have collected up to now.

In the summer of 1981 I was asked by a golfing colleague, Dave Pemberton, to give my opinion on what had caused some flattened circles which had appeared in a local field. Apparently they had been seen there and elsewhere on numerous occasions in previous years. I had no idea what he was talking about and was unable to relate his descriptions to any well-known phenomena, but my interest was aroused and I visited the site the same day.

When I arrived at Cheesefoot Head, I found myself looking down at a field in a natural amphitheatre. Below me were three large flattened circles in a line. The impact they had on me was profound. I am not sure how long I stood there, staring down at them, but while I gazed I remember considering all the possible ways they could have been created.

Later I decided I ought to share the experience with other people, so I contacted several national papers, along with the BBC and ITN. Local papers jumped on the bandwagon as soon as they could get the story into print. Photographs of the circles appeared alongside speculative accounts of how they were caused. Inevitably, little green men were blamed.

Over the next few years I was contacted by numerous interested enquirers and visited by interviewers and reporters. Everyone wanted to know how the circles, which were now being found elsewhere in the country, were created. At that time there was little recorded information about them. The Australian 'Tully' circles of 1966 were the best documented, but nowhere had continuous, in-depth study been carried out.

In 1985 Busty Taylor, who has a private pilot's licence, was staggered to see a circle group in a field of wheat far below him. He was so amazed at what he saw that he nearly fell out. Then he heard of the circles in the Cheesefoot Head area, where he and I met soon after when I was studying that year's group in the Punch Bowl. We found we got on well together, an affinity which has strengthened with time. Colin Andrews contacted me in 1983, and it soon became clear that we were on the same wavelength and that further discussion would be useful. Don Tuersly, who describes himself as an interested observer, has accompanied us on some of our site sorties.

The subjects of our investigations, circles and rings found depressed in growing crops, have puzzled and fascinated large numbers of people in a great many countries. The circles are areas in which a crop is laid down to form the swirled, flattened floor of a circular shape, like a shallow, straight-sided dish. In a ring, the crop has been depressed to form a circular path. Sometimes a circle is surrounded by a ring, but various other configurations have also been found. As far as we know, in Britain this phenomenon was originally confined to southern England, but recently a group of circles in Leicestershire and a single circle in Oxfordshire have appeared. They have also been found in other countries.

The circles and rings present numerous inconsistencies in their formation, 'floor' details, weather conditions, moon phases, location and other parameters. As soon as we think we have solved one peculiarity, the next circle displays an inexplicable variation, as if to say, 'What do you make of it now?'

We have included a large number of photographs to help you, the reader, feel as though you are visiting the site yourself. You may see something new: some detail, a possible force-field connection, have a fresh thought or notice a similarity to some other phenomenon. We ourselves have taken innumerable photographs, travelled hundreds of miles, interviewed a large number of people and spent many hours in discussion with other people and amongst ourselves trying to understand these strange occurrences.

Chapter One consists of photographs and descriptions of rings and circles to make the complexity and extent of the subject clear. The large number of locations, site groups and formations is very surprising. Some of the detail within the rings and circles is even more surprising, and Chapters Two and Three discuss this in depth. It is the detail which presents a mystery within a mystery. Standing in a circle, completely baffled, we have often remarked, 'However was that floor pattern created?' The rest of the book covers the history of the circles, our measuring techniques, their appearance in other parts of the world, and various theories and conclusions.

As a group, coming from different walks of life and occupations, we are drawn together by our mutual interest in the rings and circles, but, as might be expected, we all have different ideas about what causes them. We find this is a very healthy state of affairs in which to conduct our investigations. It prevents narrow-mindedness, however unintentional, and keeps us open-minded.

Colin is chief electrical engineer with the Test Valley Borough Council, and is responsible for all electrical installations in west Hampshire and for formulating, implementing and maintaining the arrangements for any civil emergency incident. Busty is a qualified driving instructor with his own driving school. His private pilot's licence has meant the rest of the group can fly with him, and the photographs taken from the air are some of the most awe-inspiring in the book. Don is a retired sign-writer who is interested in the controversy surrounding UFOs. As for me, I am a retired electro-mechanical design engineer. Recently I have been working on the design of electro-chemical analytical instrumentation, as used in nuclear power plants, and the design of wind-mills to produce electricity. Earlier I spent seven years in the Australian desert, working first for the British Missile Testing Range and then for NASA at a deep-space tracking station. This aroused my interest in unusual or inexplic-able phenomena.

After reading all the evidence people may wonder whether the circles have been created by hoaxers. Based on considerable experience, we think there are certain aspects of a 'true' circle that could never be produced by a machine or manually. And this is the heart of the problem. How were these circles created in such a manner? The question is extremely hard to answer, particularly bearing in mind the following facts: the circles and rings are usually created at night; the stems of the affected crops are usually undamaged in any way; they are flattened when green, green and wet, green and dry, ripe and wet, ripe and dry; they appear in dry or wet weather, warm or cold, it makes no difference.

The question 'Why?' is another problem altogether. If we could answer that, we could probably provide answers for many other phenomena which have parallels with this subject. And if you feel that, having studied the evidence gathered here, you have an answer which satisfies all the parameters described, you will have solved one of the world's myster-ies. For *Circular Evidence* is just that – evidence of mysterious occurrences which we have visited, stood in, flown over, photo-graphed, measured, examined in careful detail and presented for your curiosity or maybe your profound study. We have set out to prove no more than that genuine, anomalous circles and rings exist and probably have done for count-less years.

What force creates this strange phe-nomenon?

We realise the hoax is the favourite explana-tion for those who debunk the unknown, so we will leave it to the reader to attempt a rational explanation after considering all the complica-tions, inconsistencies and non-conformance with general science. It probably does not require genius, just straightforward thinking.

PAT DELGADO
SEPTEMBER 1988

CHAPTER ONE

---◇---

THE EVIDENCE

(All descriptions by Colin Andrews unless otherwise stated)

Plate 1.

HEADBOURNE WORTHY, 1978

Ian Stevens was driving his combine harvester in a field of wheat near Headbourne Worthy in Hampshire during the summer of 1978 when he came across a very large circular area some way into the field. Inside the circle the crop was pressed firmly to the ground and swirled around in a clockwise direction. Plants were splayed out from the centre towards the perimeter. He stopped his machine, then noticed four smaller circles near by. They appeared to be regularly spaced around the large central circle. The formation looked like the five dots found on a dice.

Intrigued by his discovery, he steered his combine harvester round the set of circles so as not to destroy them, then continued harvesting till lunchtime. When he got home he urged his wife and daughters to take a look.

Plate 1 was taken by Mr Stevens inside the central circle, with his family leaving the field in the background. The A34 from Southampton to the Midlands runs behind the hedgerow (centre, left to right). Many more mysterious circles have been found along this stretch of road north of Winchester. Mr Stevens himself has seen a number of them, but this was the first time he had seen them in such a configuration.

He first saw a single circle in 1976, on the western side of the A34, from where it could easily be seen. About 6·5 metres in diameter, it consisted of plants pressed to the ground and swirled around, like his latest discovery. It appeared about seven weeks before a Mrs Bowles had seen a UFO just down the road. Circles have appeared in the summer around here ever since.

Plate 2.

LITCHFIELD, 1981

Just north of Litchfield, a village on the A34, is an ancient burial site called Seven Barrows. Two more circles were found here in the summer of 1981 in a field alongside two prominent burial mounds. The circles were approximately 15 metres in diameter and one is known to have been swirled in a clockwise direction. The angle formed by a line running through the two circle centres was the same as that formed by a line running through the two mounds a few metres to the north. The diameter of the mounds was also equal to that of the circles, which had formed overnight. From the air they would have looked like a duplicated pair of circles side by side.

The site was interesting for a number of reasons. The two tumuli are prominent and untouched. They lie to the west of the road, almost underneath the 400,000 volts super grid electricity cables which form part of the UK national grid network. A disused railway runs north to south parallel to the road. The construction of both road and railway resulted in most of the Seven Barrows being destroyed.

The two circles formed mysteriously, despite being near a very busy road. As usual, nobody saw anything out of the ordinary happening. Thousands of travellers could see the circles as they passed by in their cars, buses and lorries, between the Midlands and the south, but what caused them was not discovered.

The locals, particularly Geoff Thompson, the farmer, and his employees, recognised that something odd was happening. They studied the details inside the two large areas and were at a loss to explain the cause. They had not seen anything like it before. Tim Martin, a student on the farm, told me he thought they could have been caused by army helicopters, which frequently fly low over the area. Mr Martin took Plate 2 from the A34, looking west across one of the circles. He said, 'I have heard of similar circles appearing near military areas at Warminster and on Salisbury Plain in Wiltshire and at Wantage in Oxfordshire.' We also knew of others in military areas, but could draw no conclusion from that at this stage.

The mystery deepened in 1982 when two more circles appeared at the same place, within a few metres of the 1981 location.

Plate 3.

THE PUNCH BOWL, CHEESEFOOT HEAD, *1981*

This site was the first to arouse national interest in the circles. Plate 3 shows the group of three which were widely publicised on TV and radio and in the national press. They were the first circles I had ever seen at close range and the impact they had on me was sensational.

I remember being asked what I thought about them before I had seen them. They turned out to be far more beautiful and precise than I had imagined from the description I had been given. The centre circle was 17 metres in diameter and the smaller ones were 8·5 metres. They were all swirled clockwise and had apparently been formed in late July. I was so impressed by what I saw that I decided the public should know about them.

Since then, my investigations have brought to light many other circles in various parts of the world, including England, which had been found before 1981. However, to these circles goes the honour of starting the flood of interest and the exchange of information about this phenomenon.

It is very important to note that there are no well-defined tractor 'tramlines' in this view of the field because at that time crop spraying was kept to a minimum. In later years tractors were used to spray the crops more intensively.

Plate 4.

LITCHFIELD,
1982

In 1982, for the second year in succession, two circular areas of flattened corn appeared in a field next to the Seven Barrows. In both of them clockwise-swirled floor patterns could be seen. One was almost 12 metres in diameter and the other a little smaller at 9 metres. The plants around the perimeter stood upright and were totally unaffected.

Derrick Goddard took Plate 4 from the western side of the A34. Mr Goddard is a manager with an agricultural engineering maintenance company and lives near Winchester. He repairs machinery on farms throughout central southern England, and has reported circle discoveries to Pat over a number of years. It was on his travels that he discovered the two depressions in a wheat field. Pat was the only person to show any interest in these marks, apart from Derrick's friends, Petronel and Martin Payne, who work on Longwood Estate near Cheesefoot Head. They were very interested in what Derrick had to say because they knew that similar circles had been found on the estate. They had talked to employees who had seen them in 1979, and others, themselves included, had seen several in the area since then.

Derrick returned to Litchfield with three colleagues. Looking closer, they found a small area of flattened corn to the north of the two circles at the edge of the field, next to the field with the two burial mounds. The semi-circular area looked as though it might have been the edge of a third circle, two-thirds of which was not visible because it was formed over a hedgerow. The grass alongside was unaffected, as was the hedge itself.

Apart from the ancient burial mounds, the location has historic significance for other reasons. Geoffrey de Havilland made his first flight on 10 September 1910 a few metres north of where the circles were discovered, and 700 metres further north is the burial place of the Fifth Earl of Caernarvon, who died in 1923 shortly after discovering Tutankhamun's tomb. And the ancient wayfarers' track, from the east to western England, crosses between the two fields.

I interviewed Geoff Thompson, the farmer, during 1986. He told me he had not seen circles like this before 1981 or since 1982 and was at a loss to know how they were formed.

Before leaving the village, I met Albert Bunce, a retired farm worker. He claimed to have seen two circles in 1982 and that many accidents had happened near that spot since. This I can confirm, having spent fifteen years with the Hampshire Fire Brigade in that area.

Plate 5.

HEADBOURNE WORTHY, 1982

At Headbourne Worthy, twelve miles south of Litchfield, unexplained circles appear most years. Tim Brown, the farmer, said, 'Nobody ever sees them formed, we usually come across them when we are working in the fields.'

Ian Stevens, who found the circle in 1978 (see pages 16–17), is a farmhand on Mr Brown's farm. He was the first to see this 11·5-metre-diameter single circle. The floor area within the circle was flattened and swirled around in a clockwise direction. Again, it was close to the busy A34, but this time on the north side. Plate 5 was taken by Mr Stevens from his combine harvester. Mr Stevens recalls being surprised that no plants were damaged even though they had been pushed firmly to the ground in a very tight swirled pattern. 'It was just like something had landed in the cornfield from the air and gone back up again,' he said. 'I don't know what to make of these things.'

Plate 6.

CORHAMPTON, 1984

Two research organisations carried out a survey of cereal farmers in Hampshire and Wiltshire. The survey forms submitted by farmers who indicated they had found circles on their land were passed to me for investigation.

One form came from Charles Hall of Corhampton Lane Farm in southern Hampshire. Since this was a new site, I was keen to visit Mr Hall as soon as farm work would allow. When I arrived, his employee, Pat Lanham, was with him. Mr Hall told me that his wife, Barbara, had been the first to spot the circle when she was out riding during the first week of August 1984. She collected her camera and returned immediately to the field where Pat Lanham was harvesting. Pat had stopped near by and was staring at the circle in disbelief.

Mrs Hall climbed on to Pat's harvester and took several photographs. One of these, Plate 7, shows her son standing in the 25·6-metre-diameter circle. The tightly swirled floor flowed in a clockwise direction.

I asked the farmer if there were any water-collecting ponds or reservoirs near the circle. 'Well, yes, there is an old dew pond within five or six metres of the edge of where the circle was found. We filled it in just a couple of years ago, but it was there in 1984,' he replied.

Amongst other possible connections, I had noticed that a large number of sites had dew ponds, reservoirs and underground water tanks close to the circles. Mr Hall mentioned that there was a second dew pond, which was still active in the north of his estate.

Only weeks after my interview with him, a second circle was found in the north of Mr Hall's estate, within 100 metres of his other dew pond.

Plate 8.

MATTERLY FARM, GANDER DOWN, 1985

One Saturday morning my wife and I were driving along the A31 from Alresford to Winchester. It was a beautiful, sunny morning and we were going up the hill beyond Matterly Farm. The fields on either side were planted with cereal crops. As we neared the brow of the hill, I spotted a set of five circles in the dice-dot formation in a field directly behind a lay-by. I swung in to gaze with wonder at the circles, which were less than twelve hours old. I could be sure of that because I had checked all these fields' the previous evening during one of my daily sorties.

While I was photographing the circles, a number of people stopped to see what I was so interested in. Later I rang an independent television company, who were immediately interested in my description. The following morning a reporter and camera crew came with me to the field and took some lengthy shots of the circles.

The large central circle measured 16 metres in diameter and the four outlying smaller ones were all 3·2 metres. All five circles were flattened and swirled clockwise, as can be seen in Plate 8. Their edges were sharp with very little gradation and none of the flattened stems appeared to be damaged. The group remained almost in the condition they were found in for about seven weeks, after which the field was harvested. Some tracks appeared as sightseers stupidly walked across the field instead of using the tramlines, much to the farmer's understandable annoyance.

Some weeks after the field had been harvested and ploughed, five green circles could be seen where the grain had fallen from the ripened heads of flattened wheat and had sprouted.

When the circles were first formed, the crop was quite green, but the flattened stems made no attempt to grow vertically again. They continued to grow horizontally and ripen, as in all other cases of this phenomenon. Whatever the crop, it seems to be induced to continue growing horizontally.

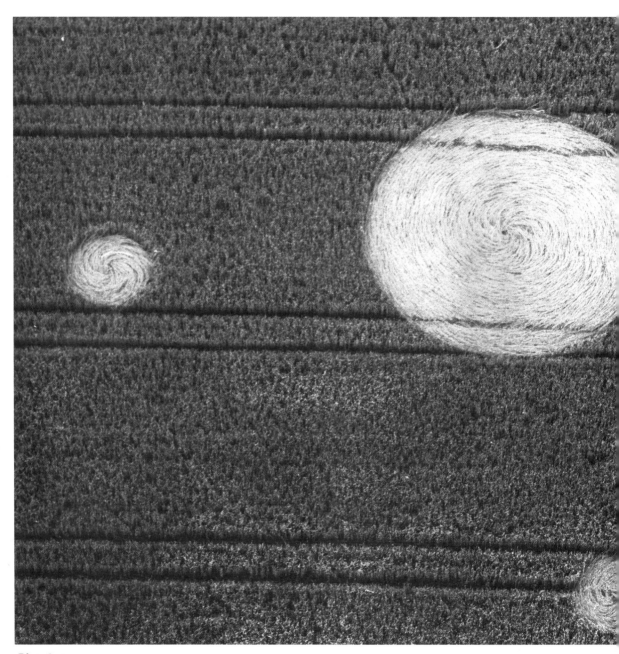

Plate 9.

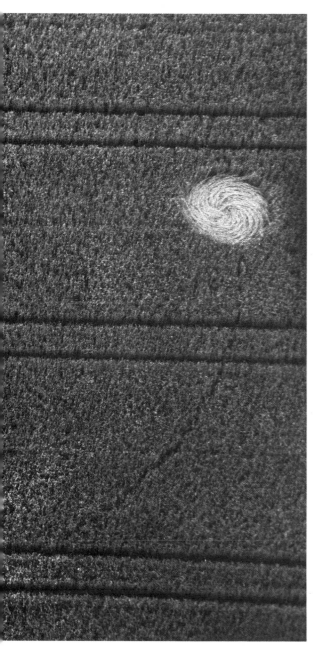

GOODWORTH CLATFORD, 1985

Busty Taylor, who is a light aircraft pilot and a member of our research group, had never seen or heard of the circle phenomenon until Saturday, 3 August 1985. He was in a light aircraft which had taken off from the famous racing circuit at Thruxton, west of Andover in Hampshire, on a course towards Andover and then south towards Stockbridge. Just before making visual contact with Stockbridge, he noticed a fantastic formation of five perfectly formed circles clearly depressed into a corn field below him. Busty and the two passengers could not account for their find, but noted the approximate position of the aircraft. 'The circles were like five Catherine wheels looking up at us,' said Busty. That night he was sleepless as his mind worked overtime trying to figure out how they had been formed.

The following morning he was back at the airfield to try to relocate the five circles and photograph them. A trip which should only have taken him five minutes took half an hour because the remote site was more difficult to find than he had anticipated. Taking photographs from 152 metres, he could now see the circles in great detail. The spiking around the perimeter of each circle was very impressive, as was the spiral pattern twisting back into the centre. He was now even more baffled.

The following afternoon he drove to the field with a reporter and photographer from the *Andover Advertiser*. They had barely reached one of the smaller satellites, taking great care not to damage the crop, when an Army helicopter appeared from the direction of the ancient hill fort of Danebury Ring. It headed directly towards them, then hovered overhead for some minutes. As it moved slowly away, Busty, who was looking down into the circle, suddenly saw a bright blue flash like that from a camera flash gun. The journalists were still looking at the helicopter, now turning in the distance to face them again, and did not see it. There have been other reports of bright flashes from the vicinity of circles, see pages 94–95.

During discussion with a senior Army officer, we were later informed of a report sent to the Ministry of Defence about a 12-metre circle in which the corn had been laid flat in a clockwise swirl. The wheat on the edge of the circle was completely upright and undamaged. Four separate smaller circles, approximately 3·6 metres in diameter, were set out in a precise square, north/south and east/west. Their centres were 43 paces from the centre of the large circle. More evidence of military interest in these unusual occurrences will be referred to later.

Plate 9 was taken from a helicopter directly overhead. Three of the satellites were joined by a circular track running through their centres. The plants along this track were not pressed to the ground like the others, but were just dipped over at their heads.

My father, Gordon Andrews, can be seen in Plate 10 inspecting the floor pattern of the large circle, close to the spot where Busty had earlier

Plate 10.

found a luminous, white, jelly-like substance which is still unidentified. It was analysed at the University of Surrey in Guildford and at the Albury Laboratories, a national testing laboratory in Surrey. Plates 11 and 12 reproduce the sample reports prepared by them. Note that, after reaching a conclusion, the University decided to test it with Fehling's solution, which gave a negative result, whereas commercial sweets and honey give a positive result.

Pat and Jack Collins, both pensioners, reported a UFO sighting to Hampshire police in the evening of 6 July 1985. They were driving over Stockbridge Down on the A272 when they saw a huge circular object standing on end, like a funfair wheel. It was stationary and hovering close to the ground over the downs. 'We were only about two hundred yards from it,' Mrs Collins told me, 'and it had lots and lots of yellowy white lights all around the edge and more lights along spokes leading into the centre of it.' They were terrified. Two police cars searched the downs, but nothing was found. The following morning the five mysterious circles were found near Alresford, see page 30, and the five at Goodworth Clatford were found later. The two circle sites and the UFO sighting form a straight line. An identical wheel-type UFO was seen by four people between circle fields near Warminster in August 1982.

 UNIVERSITY OF SURREY

Guildford Surrey   Guildford (0483) 71281   Telex 85331

Report on Sample 'SIGAP 11/8/85'

Initial examination by optical microscopy was undertaken by Dr.Moss. He did not find any evidence to indicate that it was a slime mould (Myxomycetes) as had been suspected. He noted the following :-

1) A large number of starch grains,which were identified by their reaction with iodine solution.
2) Crystals which effervesced with dilute acid, thought to be calcium carbonate.
3) The absence of slow dissolving crystals such as sugar.
4) A small number of cell structures which were thought to be of plant origin.
5) A large number of bacteria were present.

The conclusion he reached was that the sample was some kind of confectionary which had gone off,he also noted a slight smell of honey.

The sample was further examined as follows:-

1) The sample was shaken with water. It was incompletley soluble and showed some frothing.

2) A test for reducing sugars was made with Fehling's solution. This was negative. Glucose syrup used in most commercial sweets and honey give a positive reaction with this test.

ALBURY LABORATORIES LTD.

Received from yourself three samples for examination:

1. Soil in "ring"
2. Soil control sample
3. Jelly like material from "ring" centre.

These were subjected initially to a scan for radiation and no X or Gamma radiation was emitted from any of the samples.

The two soils were compared for pH and nitrate and the values obtained are given below:

	Sample 1.	Sample 2.
pH	6.7	6.8
% nitrogen	0.40	0.35

All samples were viewed in normal and ultraviolet light and no differences were observed.

Sample 3 the jelly was examined bacteriologically and found to contain normal soil flora e.g. Bacillus s.p. and coliform organisms. No distinctive or unusual features were observed.

Plate 11.   *Plate 12.*

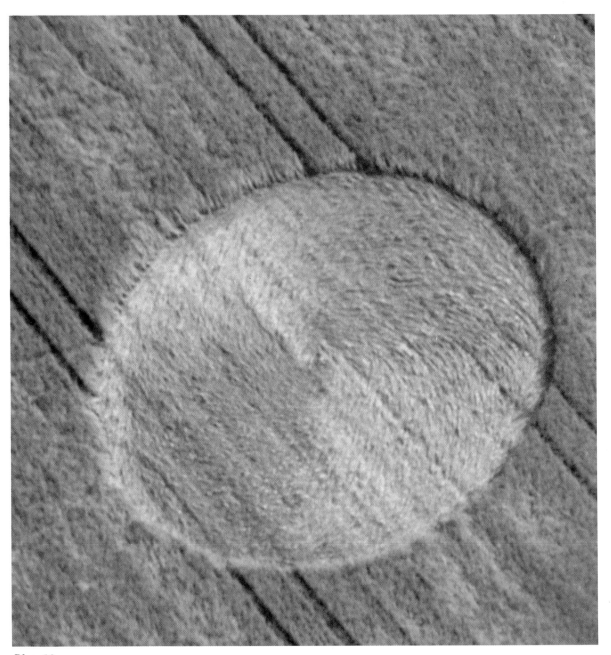

Plate 13.

Upper Farm, Headbourne Worthy, 1986

This circle was discovered by Busty Taylor during the morning of 1 August 1986. Busty had scanned the area the previous evening and it was not there. It was a case where being tuned to instant recognition helped in finding circles. Driving away from Winchester towards Andover on the B3420, this particular field offers a very shallow sloping angle, and there is only one gap of any length in the hedge through which to make a quick scan.

I arranged to meet Don Tuersly at the field entrance later. However, Don called at the farm first and asked the farmer, Simon Brown, for permission to enter the field. Mr Brown asked Don which field it was, and when told said, 'Oh, it's in that field this year, is it?' When Don asked what he meant, he replied, 'My father used to point these circles out to me when I was a lad. They have appeared in one of my fields almost every year for the last twenty-eight years that I can remember.' He did not seem to think that what he was saying was sensational, but acted as though he thought it was quite commonplace.

He gave us permission to enter the barley field and our measurements showed that the diameter was 17·4 metres and almost a true circle. The floor was flattened in a counter-clockwise swirl with the whorls showing beautifully. The central swirl was 100 millimetres in diameter and was constructed to look like a pot or vase, 200 millimetres deep. The edges were sharp with no serrations, but there were very slight 'gap-seeking' incursions where the tramlines entered and left the circle.

It was a perfect example of a large single circle with the stems packed hard down to the ground without damage. The field sloped gently down from north-east to south-west with rolling countryside all around. The circle was barely visible from any point around the field. The only good view was from the air when Busty flew us over the site a few days later to take some photographs, one of which can be seen in Plate 13. Plate 14 was taken on 8 October and shows the circles some weeks after harvesting.

Pat Delgado

I arranged to meet a colleague in Down Farm Lane near Headbourne Worthy at about 12.45 pm on Saturday 16 August. The lane runs parallel to and south of the A34 and is a turning off Three Maids Hill roundabout at its junction with the B3420. It was a warm, sunny day with one or two small white cumulus clouds moving slowly across the sky from the south-west. My car was facing south-east with the driver's window wound down. It was 12.25 pm when, looking up into the blue sky, I caught sight of a small, grey-coloured disc-shaped object hovering high up in the sky, between myself and Winchester. The object darted at high speed, and in one jerking movement, to the south-west. It was obscured by a small cloud moving in the opposite direction. I stared at the cloud, knowing it had to emerge, but it didn't. After a few minutes it was obviously not going to appear again. Had I missed its emergence by watching the cloud, or did the object remain obscured? The sighting troubled me greatly.

Plate 14.

Plate 15.

CHEESEFOOT HEAD, 1986

Plate 15 was taken during one of our aerial reconnaissances on 23 August. In the centre are a circle and ring which appeared in the Punch Bowl at Cheesefoot Head of 5 July. The smaller indentation to its left was hoaxed by youngsters, and top right is the beautiful formation which appeared on 14 August and is the subject of this report. Plate 16 is a close up taken over it.

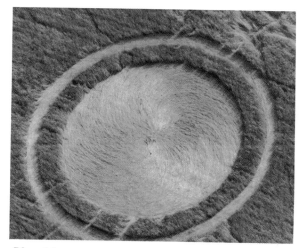

Plate 16.

We try to keep prime sites under daily observation between us. It was my turn to check the Punch Bowl on 13 August, and when I left the field at 10 pm there were two circles. At 10 the following morning Martin Payne was travelling along the A272 and saw three indentations.

The new circle had one of the most impressive aerodynamic swirled-floor patterns we had ever seen. It was 19 metres in diameter and swirled clockwise. A band of standing wheat 2·4 metres wide separated it from a flattened ring 1·3 metres wide, which swirled anti-clockwise. We had previously seen several interesting spurs pushing out of rings at around 120° (magnetic), and here too at about 120° was a swathe fanning out into an arc. It was 15 centimetres wide where it left the ring and 2 metres wide at its furthest point, where it abruptly ended 4 metres away. The beautiful floor patterns were pressed hard to the ground and great vertical pressure appeared to be responsible. Strangely, a small group of plants remained standing which had had their heads removed by mice at some stage. There must be a clue here to the energy field at work. It may be that the removal of their heads reduced their resistance to the force and saved them from being flattened.

Plate 17.

CHILDREY, WANTAGE, 1986

As a farm mechanical engineer, Derrick Goddard travels many miles around southern England. As mentioned earlier, he had seen the mysterious circles pressed into cornfields, but had not been able to arouse any interest in friends and colleagues when he discussed his discoveries.

In early September 1986 Derrick was on his way to a customer along the busy B4507 near Wantage, Oxfordshire, when he thought he could see an indentation in a field of wheat. The mark was close to the road and turned out to be a large circle surrounded by a ring. He had never seen this formation before and was stunned by its symmetry. He had read Pat's articles about circles in local newspapers and so contacted him when he returned home. Pat was unable to go to Oxfordshire immediately so Busty and I set off early on 3 September.

Although we had been given reasonable directions, I found nothing. The reason why became clear the next day when I returned to find the field had been almost completely harvested. I parked my car by the edge of the field and walked across the lines of straw to the farmer, James Matthews, who was finishing the western side. He was very keen to find out what was causing these formations. He told me he had found the damage early in the morning on 26 July. It hadn't been there at dusk the previous evening. He said, 'I phoned Wantage police station because I was at a complete loss to know how such a thing had been created. My friends took some photographs shortly before the police arrived.' Plate 17 is one of these photographs; it shows a pathway pushing out of the ring. He added, 'The police measured the whole thing and told me, before they left, to expect to find the marks again in the future.' We could not find out what led the police to say that, but it is certainly borne out by our experience.

The circle floor had a radius of 6·1 metres and consisted of clockwise swirled plants pressed to the ground. Outside this the direction of flow reversed 180° in a separate band 1·8 metres wide, but this could only be seen on close inspection. The whole flattened area averaged 16 metres in diameter. The standing band of plants was 2·2 metres wide and the Saturn-type ring was 1·1 metres wide and consisted of plants swirled anti-clockwise and pressed to the ground. The formation also contained the most unusual feature we had so far seen. At 120° (magnetic) a pathway 1·2 metres wide pushed out of the circle. The plants were pushed straight out for 5 metres to where the walls tapered in to form an arrow head. At the point of the arrow was a circular hole 0·35 metre in diameter. It was about 0·23 metres deep and dome or bowl shaped. 'The sides were so smooth', Mr Matthews said, 'that they must have been cut and soil removed by some kind of tool.' The soil that had been removed was nowhere to be seen, which had puzzled the police. Mr Matthews added, 'A neighbour told me some experts say that these things are being caused by whirlwinds. I'm positive a whirlwind could not do this, I've seen plenty in my time.'

An extraordinary series of events happened at my home shortly after this. Too much happened to report here fully, but since the events were paranormal it seems appropriate to give an outline. When I returned home I deposited a soil sample from the hole in an office specially prepared for this research. The room has very intricate burglar alarm systems, as do the house and the perimeter of the property. I locked the office door and set the alarms. Minutes later an infra-red detector had sensed movement inside the office and activated one of the alarms. I did not find this disturbing, though I could find no reason, because I design alarm systems and know their shortcomings. But worse was to come.

At 4.15 the following morning another alarm sounded. This time it was the system protecting the perimeter. I found a time clock, which is mains voltage operated, had stopped at 4.15 am and was now faulty. I found no cause for this, but the next day the clock was working again. Again, several nights later, it stopped working at 4.15 am. And the office alarm, which is separate from the perimeter system, was also sounding, and had been activated at 4.15 am. My wife guarded the side door to the house as I walked to the office and unlocked the door. A microwave detector had been activated and the battery-operated wall clock had stopped at 4.15 am.

After further occurrences around the house in the following weeks, my wife suggested I end my research work. She is not interested in the paranormal or the fringes of science, but even she thought that some entity appeared to be behind these events which might be connected with the mysterious circles.

I have discussed the above at length with Helen Tennant of the British Psychical Research Society and Professor Archie Roy of Glasgow University. They agree that no rational explanation can be found.

Longwood Estate, 1986

Plate 18.

On 23 August Busty, his son Nigel, and I were on one of our regular reconnaissance flights. It was a fine morning with good visibility and as we approached Cheesefoot Head from the north we could see the two formations and their hoaxed companion in the Punch Bowl. Although we had already investigated them fully, we took more photographs. As we climbed away to the south-east, Busty said, 'All we want now is to find all the formations we have seen to date wrapped into one, like the Celtic cross.' Next morning Busty was flying over the area again with Omar Fowler of the Surrey Investigation Group into Aerial Phenomena. At the very point where he had made this remark, the astonished pair saw below them all the formations seen to date wrapped into one, forming the Celtic cross. It had appeared since the previous flight thirty-one hours earlier and the intervening night had been very wet with heavy rain (see Plates 18 and 19).

When our research group gathered there two and a half hours later we found a large circular area of flattened green wheat, 12·9 metres in diameter with the stems spiralling clockwise. A band of standing plants, 1·8 metres wide, separated it from a ring of anti-clockwise swirled plants 1·5 metres wide. Four smaller circles were positioned at the

Plate 19.

cardinal compass points, forming a square or cross. All four were swirled clockwise and 3·6 metres in diameter. A narrow, partly formed ring cut through the centres of three of the satellites and the whole formation was 47·5 metres in diameter.

Plants in the tramlines grow more slowly and so are shorter and less ripe. These remain standing and are unaffected by the energy field which flattened the other plants to the ground.

Some families were picnicking near by totally unaware of the markings in the field next to them. Only when you are familiar with the phenomenon do you recognise the indentations from a distance, showing as a darkened area in the corn. This formation, huge as it was, could only just be seen from the A272, Winchester to Petersfield, road, as it approached Cheesefoot Head. Maurice Botting, farm manager at Longwood Estate, told us later of a large underground water tank on the edge of the formation.

Plate 20.

South Wonston,
1987

Here a single circle 14·42 metres in diameter appeared in rape crop. It had a tight, clockwise-swirled floor pattern and there was evidence of mechanical contact in the serration marks found on some leaves and stems. All the plants inside the circle were pressed firmly to the ground. Some tooth-shaped swathes around the edge were spiked into the standing wall. The circle was 23·5 metres equidistant from a public highway and a farm track, and was formed on 8 May 1987 and first visited by us on Sunday 10 May.

The Saturday before all had been quiet, until the telephone rang. It was Pat to say the first circle had appeared early this year and had been found by Derrick Goddard's son between South Wonston and Sutton Scotney. He had been cycling home along the old A34 when he spotted a darkened area in a yellow rape field, just south of the gantry reservoir at South Wonston. When he reached home he told his father and they both went back in Derrick's van. It was late and the light was fading, but Derrick could just see that the darkened indentation was a circle.

We phoned the farmer, David Stainer, to ask his permission to carry out a full scientific site study. He agreed, so we gathered at 8.30 the next morning outside the field. This single,

Plate 21.

Plate 22.

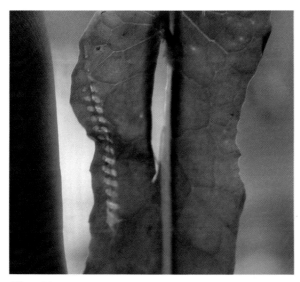

Plate 23.

clockwise swirled circle was the first we knew of in a field of rape. Plate 20 shows the position of the field north of the A34. Plate 21 shows the team at work in the circle. Both photographs were taken from the air by Busty.

Rape plants have a substantial stem structure which is brittle and snaps rather than bending when pushed over. But most of the plants in the indentation were not broken. Don noticed that they had been pushed down firmly to the ground, as if steamed into position, causing the stems to bend just above the ground at a point where they are hard and fibrous. Although they were close to the ground too, the delicate yellow flowers were intact. One or two stems and leaves had serration marks on them, and each one was approximately 1 millimetre wide with a spacing of about 2 millimetres between marks. Rape plants bruise easily when touched, and so gave us our best opportunity so far to obtain clues about mechanical contact. The markings found here (see Plates 22 and 23) were consistent with those found later in another field of rape at Corhampton.

Don and Peggy used to live at South Wonston and know the farmer, Mr Stainer, well. We were standing on the edge of the field when he drove by with his family on their way to work. He stopped and said, 'My wife saw it on Friday afternoon. It's in the same place as one we had in corn back in August 1983.'

On 8 June I asked my father, Gordon, to help me take two measurements at South Wonston which had not been taken on earlier visits. My mother, Elsie, and the dog came too. We parked at the end of the track down to the farm. Elsie stayed in the car with the dog while we went straight to the circle.

My father held the end of the tape and stood in a circle while I walked with the other end towards the hedgerow on the western side of the field. As I approached the hedge, about 10 metres from my father, I saw a bright flash of light. I flinched and turned to face him. He was looking pensive. He had heard a loud crackling noise at the same time. We returned to the car, where Elsie and the dog were shut inside with the windows closed. It was parked a good 25 metres from the circle, but she too had heard the crackling noise. 'I looked around at the rear seat, thinking the dog was walking on a crinkly paper bag, but there was nothing there,' she said.

Plate 24 was taken on Saturday 1 August, showing the circle three months after it appeared.

Plate 24.

CORHAMPTON, 1987

This was another single circle, 12·1 metres in diameter in rape crop. It had a tight, clockwise-swirled floor pattern and there were serration marks identical to those found at South Wonston. The plants were pressed firmly to the ground, with new weed growth inside the circular area. The circle was roughly equidistant from a minor road and a farm track. It is thought to have been formed on 1 June 1987, and was first visited by us on 15 July.

We had first visited Corhampton Down Farm on 21 May, investigating a circle said to have been found on the farm in 1984 (see Plate 7). During the interview, we pointed out to Colin Hall, the farmer, how unusual it was for only one circle to appear at sites we had investigated. So we were not surprised when he phoned on 15 July to say that another circle had been found, which he thought was about six weeks old.

Mr Hall told us that it had appeared within 100 metres of a dew pond. 'We only have two dew ponds and now we have found a circular area next to both.'

The site is just south of a Roman settlement on high ground and overlooking Southampton and the Isle of Wight some miles to the south. Plate 25 shows the circle as we found it within two hours of Mr Hall's call. Busty can be seen

Plate 25.

with some of the research equipment, and the dew pond is behind the hedge at the top of the picture.

The circle was in a remote location on the brow of an incline, so it could only have been seen from the air or from very close range within the field. Measurements confirmed that it was unusual in being a perfect circle, which most are not.

What kind of force could be responsible for such an area of flattened plants?

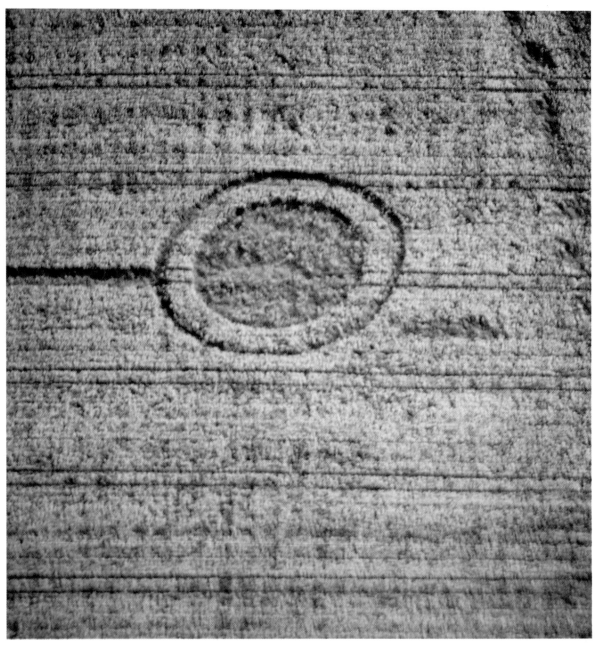

Plate 26.

WHITEPARISH, 1987

A single circle, 15·38 metres in diameter, was found in barley crop inside an annular ring 1·18 metres wide. The standing barley between the circle and ring was 2·68 metres wide. The circle was swirled in a tight clockwise floor pattern and the ring was counter-clockwise. A 2·15-metre-wide swathe of plants, flattened outwards, formed a 14·01 metre spur out of the ring at 126° (magnetic). Many plants in the formation had begun to lift back up. The pattern was formed on about 1 June 1987.

I first visited the site on 9 July. I had arranged to meet Dr Terence Meaden less than a quarter of a mile west of Pepperbox Hill, an archaeological site and well-known beauty spot near Whiteparish because he had been told by Paul Fuller of an odd marking in a field east of Salisbury.

The three of us met up and made some enquiries about who owned the land, but no one could help, so, with the evening light fading, we entered the field from a small road. We spent a few minutes looking at the formation.

As we left the field we met the farmer, Gordon Sparkes. 'Who the hell are you? What are you doing in my field?' he exploded. We explained that we were a team of engineers and scientists researching phenomena of this kind throughout southern England. Mr Sparkes was not impressed and reversed his car up the lane at revs previously considered impossible. As the blue smoke cleared, we could see that he was blocking our exit. He jumped out and locked the door, then snatched my notebook and left on foot, leaving us shivering beside our trapped cars.

Some thirty minutes later he returned with a police officer and gradually it became clear why he had been so angry about circles and our presence in his field. He had farmed this land for twenty-nine years and during that time had found many similar circle marks. He was clearly convinced that they were formed by people and was desperate to catch those responsible.

PC Anderson, with great skill, calmed Mr Sparkes down, and encouraged us to explain once more the purpose of our visit. After long discussion Mr Sparkes, now in control, said, 'I have seen some kind of pod marks in the field opposite. They have been perfectly formed.' This was useful information which we presumed referred to quintuplet sets.

Mysterious circles had appeared in this area well before Mr Sparkes's time. My brother-in-law, Evan Scurclock, remembers finding several in a neighbour's field. 'We used to find them close to Pepperbox Hill, near the spot where an aeroplane crashed, around 1936 to 1940,' he said.

Interestingly, there is a military presence at this location. The escarpment overlooking this site has been used as a store from the time circles were first found. The tunnels running deep into the hillside are now used for the storage of nuclear warheads. We were reminded of this during the winter of 1986 when a specially constructed nuclear container vehicle crashed in the lane running alongside the field. A convoy of military vehicles had been escorting it to the store at West Dean when it overturned into a field. It was widely reported in the international press and Russian spy satellites were manoeuvred into position to observe the operation.

Few new clues about the circles could be found. There is a large reservoir near by; an escarpment rises up to the east; there are archaeological sites to the south and east of the field, and military establishments near by. These factors are in common with many other sites. The spur of plants pushing outwards and slightly off centre was unusual: we had seen a shorter version at Childrey in 1986, but that had tapered in to form an arrow head with a small hole at the point. Here, at Whiteparish, the spur terminated abruptly with the flattened plants leaning up against a wall of upright, unaffected plants. There was no hole. A full site investigation might have yielded more information, but this was not possible because of the farmer's attitude. In the future we would only be able to survey this site from a distance or the air (see Plates 26 and 27).

Plate 27.

Plate 28.

LONGWOOD ESTATE,
1987

This single circle, 16 metres in diameter, in wheat had a tight, clockwise swirled pattern towards the centre which slackened rapidly towards the outer edge. Several walls of plants remained standing within the circle, totally unaffected, the largest of which was a 3-metre-long and 0·7-metre-wide band which had its curved major axis running along a very well-defined swathe, which is similar to a lightly twisted long bundle of stems curving away from the circle's centre. The circle was in the corner of a field 41·9 metres from the A272 and 40·5 metres from a farm track.

Crop spraying was taking place in the Dog-leg Field on Longwood Estate on the afternoon of 2 July. Phil, a farmhand, had filled the tanks with chemicals and set off with spraying booms down. He had not been driving long when, to his amazement, he was confronted by a large circular area flattened to the ground in front of his tractor. It looked so mysterious that he decided to steer round it (see Plates 28 and 29).

When he returned to the Estate Office, he told the manager, Maurice Botting, what he had seen. The conversation was overheard by Martin Payne, who had seen circles before in surrounding fields. Martin asked Phil whereabouts it was, and on hearing said, 'That is the same spot where the manager found five circles in July 1984 while flying over with neighbouring farmers in a helicopter.'

Martin's wife, Petronel, told us early on Wednesday, 3 July of this discovery, so we set off immediately after obtaining Mr Botting's permission to enter the field. We carried out a detailed survey that day. On the Saturday we had a long meeting with Professor Archie Roy of Glasgow University and Helen Tennant of the British Psychical Research Society, a meeting which we recorded although the tape was blank when we played it back. We then took them to various circle sites, ending up at Dog-leg Field. Professor Roy agreed with us that the lie of the plants suggested a tortional spiralling force could be responsible, which at the same time oscillated along a large number of sine wave bands from the centre outwards. Being a psychic sensitive, Helen Tennant thought the circle felt as though she should remain towards the outer rim and not explore the centre.

The circle was believed to have appeared on about 25 June, the same day that a further three circles and a ring were found in a field at Kimpton near Andover. This site was to witness some extraordinary events which would change the whole atmosphere surrounding this research.

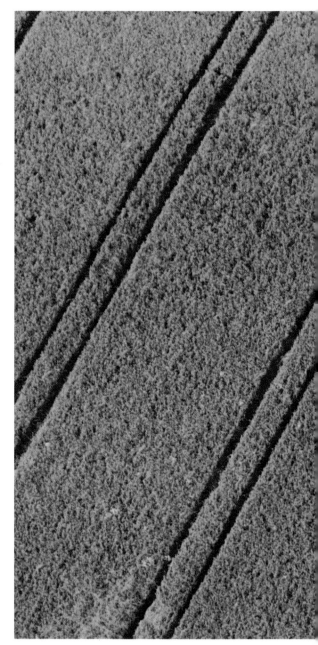

Plate 29.

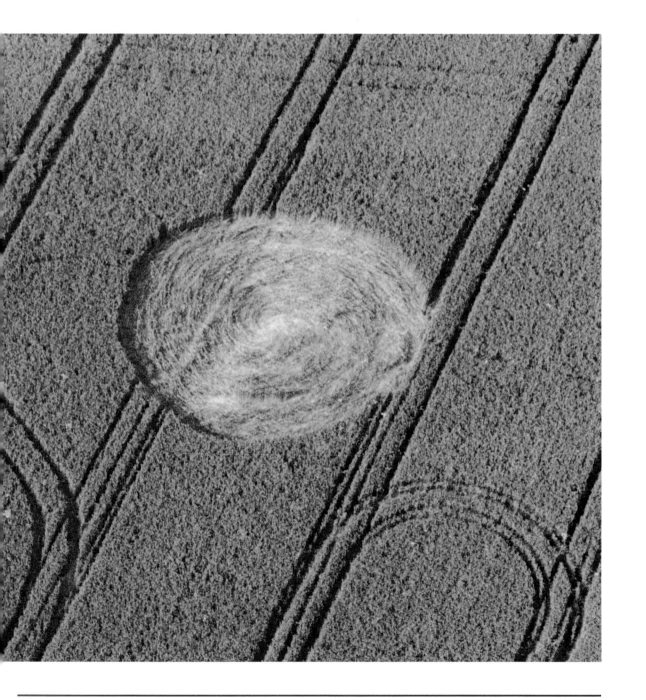

Plate 30.

KIMPTON,
1987

Kimpton is near the air approach to Thruxton airfield. As Busty was landing there one day he caught sight of a ring in a dark green field to his right (see Plates 30 and 31). He knew this area well, so could relocate it without difficulty, but he thought it was not too important because he had only seen large circles before. Some days later he mentioned it to me casually.

I decided to spend my lunch hour at Kimpton the next day, so drove over and parked outside the cemetery. I walked through the old graveyard, looking for a gap in the hedge through which I could see the two fields Busty had told me to head for. On my way back to the car to find the farmer, I met two teenage boys looking for a ball. One of them, Jamie Catt, said he had seen an orange glowing object over the field behind the cemetery on Saturday, 13 June at about 10.30 pm. Later I was approached by two elderly villagers who thought I was investigating odd noises they had heard coming from the field a few days previously.

The ring was a single oval in wheat, with a major axis of 10·19 metres and a minor axis of 8·9 metres. The plants were flattened outwards radially around the outer edge of the ring and faced inwards towards the centre on the inner edge. Between the two radial splays was a line of buckled plants. Each one was broken at the knuckle along its stem length. These collapsed plants appeared to have suffered whiplash damage, possibly caused by opposing forces meeting. The ring was 60 centimetres wide and in the remote depths of a large field which had been planted with both wheat and barley. It was 0·6 metres from the dividing line between the two crops and had almost certainly been formed on Saturday, 13 June.

When I told Mr Flambert, the farmer, about it he thought the circles I was talking about were the kind that normally appeared in his two large fields in long dry spells. An archaeologist from the near-by village of Charlton, called Max Dacre, had led a dig here some years ago and found a valuable archaeological site beneath the rings. We went to look at the new ring together.

Mr Flambert was taken aback. 'This isn't what I thought you meant,' he said. 'This isn't archaeological. I've never seen anything like this before.' I hadn't either. This was different and very odd. We were not to know that a similar, partly formed ring would be found a few metres from a circle at Westbury in

Plate 31.

Wiltshire in August 1987.

Later Pat, Don, Busty and I were joined on the site by Paul Whitehead and Steve Brown of SIGAP (Surrey Investigation Group for Aerial Phenomena) and Terence Meaden from TORRO. Paul is an active consultant for the *Flying Saucer Review* and Steve is a physicist. They were intrigued by a heart-shaped area at 143° in which the wheat was forced down and outwards in a fan-like shape, spilling anti-clockwise and clockwise into two halves of the ring, eventually completing a full circle.

During the next two days three things happened which completely changed my impression of the kinds of forces that might be responsible for the circle and ring formations.

On 29 June I visited the field alone to check a new swirled circle Mr Flambert had told me of close to the ring. Villagers had found two more half a mile away in the same field. I was recording details into my dictaphone from inside the ring when suddenly there was a black flash. I flinched and for a fraction of a second the sun was blotted out. At once I looked up into the sky. There was nothing in front of the sun. Feeling uneasy, I made my way towards the new circle. Sure enough a clockwise swirled circle 3·5 metres in diameter lay just 33 metres north of the ring.

Later that afternoon my parents and I returned with the family dog. Gordon Creighton, who edits *Flying Saucer Review*, had suggested that it might be interesting to take a dog near a circle. My father walked the dog into the field and it was pulling keenly, until it reached a point parallel with the still hidden ring when it stopped abruptly. My father coaxed it towards the centre of the ring, but

within minutes it was vomiting and became quite ill. My parents had to take it back to the car, whereupon its condition improved rapidly and it was quite well by the time they reached home some twenty minutes later.

To recap: an unusual ring and three circles had appeared in this field; an orange glowing object had been seen over the same spot; mysterious noises had come from the field; I had seen a large black flash and the dog had become ill. All these seemed to me to be good reasons to return later that evening. But I was to have the fright of my life. I walked through

Plate 32.

the graveyard and climbed through a gap in the hedge. I went a little way along the edge of the field and then turned east to climb the hill, before going about a quarter of a mile into the field. It was a fine evening and the sun was beginning to set over Salisbury Plain away on the horizon. I stopped about 10 metres from the eastern edge of the ring, facing the small circle to the north and the other two in the distance and thought, 'God, if you would only give me a clue as to how these are created.' A static electrical cracking noise started to come from a spot about 3 metres away. It grew louder, up to a pitch where I expected a bang to follow. Frightened, I looked towards the village to check my quickest route out of the field. I fought to control my panic and remained still. As suddenly as it had started, it stopped. It had lasted about six seconds, although it seemed longer. I saw nothing and nothing moved.

Arthur Shuttlewood has written of similar occurrences. This fringe phenomenon appeared to be associated with the circle/ring mystery and from now on was fully documented.

Mr and Mrs Hitchcock, who are retired, walked their Alsatian every morning on the farm track to the north of this field. For several days in the second week in June they both heard warbling, humming-like noises coming from the field. Mrs Hitchcock said it was like the noises on the television series *Dr Who*. She could not define the exact spot, but it was in the area of the field.

We continued with a 'feet on the ground' approach to our investigations, but it was becoming increasingly evident that we now had some other, more bizarre facts to consider.

Pat visited the field on the following day and Plate 32 shows the twin swirled circles, 3 metres in diameter, found there. Both circles were swirled clockwise, but some damage caused by the lads who discovered them can be seen.

CHEESEFOOT HEAD, 1987

This single circle, 15·6 metres in diameter, in wheat, had a tight, clockwise-swirled floor pattern surrounded by a single ring, 1·1 metres wide. The swirled floor in the ring was counter-clockwise and separated from the circle by a 2·6-metre band of standing plants. All the plants in the ring were pressed firmly to the ground but undamaged. There was pronounced spiking into the standing plants in the south-west quadrant. The formation was roughly in the centre of the field and was formed at night on Thursday, 9 July.

On the Friday Pat was pushed into action earlier than he would have wished. An elderly lady phoned to say she had found the forma-tion, which was 'just like Saturn'. She went on, 'I was there before 7.30 am and there it was in the famous Devil's Punch Bowl at Cheesefoot Head' (see Plates 33, 34 and 35).

Circles in various formations appear in this field most years, though this was the first in 1987. Pat had inspected the site the previous evening, and when he left at 10.30 there was nothing there. The more intensive our re-

Plate 33.

search, the better we become at timing the mysterious arrivals. A check with the Meteorological Office at Bracknell showed that weather conditions were not conducive to whirlwind or tornado formation, and even had they been, were not favourable for formations of that kind in the Bowl. Enquiries of all the major meteorological offices in Europe have supported the view that, by their nature, most formations rule out the possibility of meteorological causes.

The farmer, Mr Bruce, was in the United States, so Pat left a message to say he hoped he had permission to check the field and would take care during his visit. Mr Bruce is unhappy with our continued interest in his field, and with that of the public, who park constantly on Cheesefoot Head to view the sight below.

Of as much interest as the ring and circle was a small darkened hole surrounded by a lighter area some 30 metres away. It appeared to be the beginnings of a circle, which for some reason had not developed.

Later inspection by Pat and I, with Mr Bruce, confirmed our earlier impressions. Here was refutation for the sceptics who believed the hoax theory. The affected area was well away from any easy access. A long and careful walk along a tractor tramline without the help of lights would have been necessary the previous night. There were no tracks in the area concerned. Although Mr Bruce had yielded to Pat's pleading to be allowed to look at his field at close range, when we finally climbed back over the fence he made it clear he did not want to see us again. 'It is not you I am against, it is what you represent I do not like.' We wondered what he thought we represented, but it was not the time to debate the matter. Mr Bruce has since strengthened his fencing to keep out the public and instructed his staff to deal firmly with trespassers.

On the night of 9 July Martin and Petronel Payne were returning home past the Punch Bowl when they saw a row of bright lights above the trees on the far side of the Bowl. The lights seemed to be dancing along and even to be leap-frogging. They looked as though they were detached from one another. Martin and Petronel felt sure that what they had seen were not aircraft but a display of UFOs.

We took it in turns to observe the field for several nights and while we were there took infra-red photographs of the area. None of the results proved useful, though we learnt more about the heavenly bodies from Don, who is an expert. Unfortunately for Mr Bruce, more circles followed.

Plate 34.

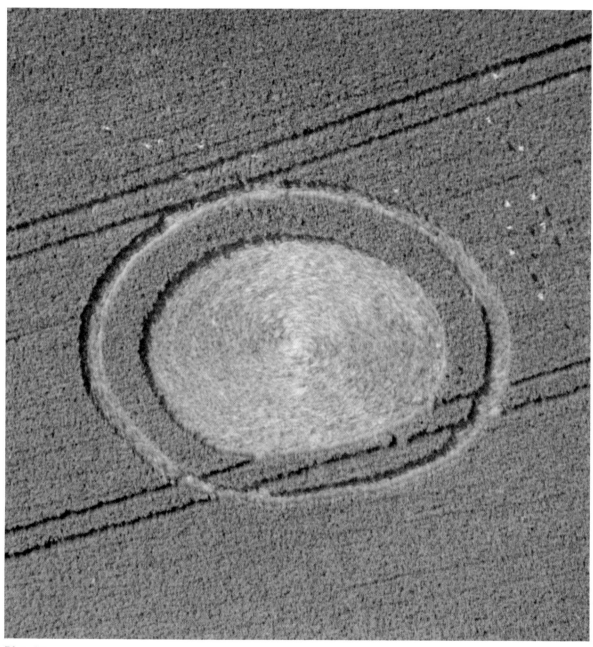

Plate 35.

Plate 36.

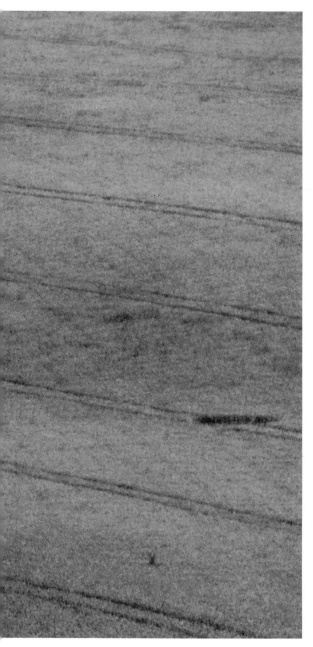

CHEESEFOOT HEAD, *1987*

At 8.30 pm on Friday, 24 July the circle and ring shown in Plates 33, 34 and 35 sat alone in the Punch Bowl, surrounded by the impressive landscape, but at 8.30 the next morning Neil Shave, a local shepherd, saw five new circles, which had appeared overnight. Plates 36 and 37 were taken that morning.

We approached Mr Bruce, the farmer, cautiously for permission to examine them. He met us at Cheesefoot Head that afternoon, with his wife Minna and their dog. He said again he was unwilling to allow us further access, but reluctantly he would allow two of us to enter the field this once. We agreed that Pat and I would examine the new circles while Don and Busty, who had brought all their equipment, would remain on the hillside and watch from a distance. Minna Bruce went with us, while Mr Bruce and his dog stayed by the road to keep other people out.

We followed one another along a tramline into the large circle. It was very impressive, the first we had seen with radial swathes – the usual swirl pattern was missing. Two sectors had plants laid slightly clockwise, and these were separated from the remaining radial areas by parting lines not unlike those in hair. The circle was 19·8 metres in diameter and the shape was within 0·1 metres of a true circle, which was quite unusual. Plate 36 shows Minna Bruce, Pat and myself investigating it and was taken by Busty on the hillside.

The four satellites were equally spaced and formed a square, but they were not north/south and west/east, which again was unusual. They were 49°, 147°, 228° and 321° (true) respectively. Two of them were a little over 3 metres in diameter, while the others were approximately 1 metre larger. The plants in each were flattened and swirled clockwise. The formation was enormous – 47·7 metres in diameter. The usual narrow track arced through the centres of the satellites, forming three-quarters of a ring between them. The plants were not damaged even though great force was clearly involved. The central circle was one of the largest we had found, its major feature being the unusual radial swathes described above. Plate 37 shows Minna Bruce leaving it, and gives some idea of how large it was.

As we parted Minna said, 'It's important you continue with your work, at least until you are in a position to explain the cause'.

Plate 37.

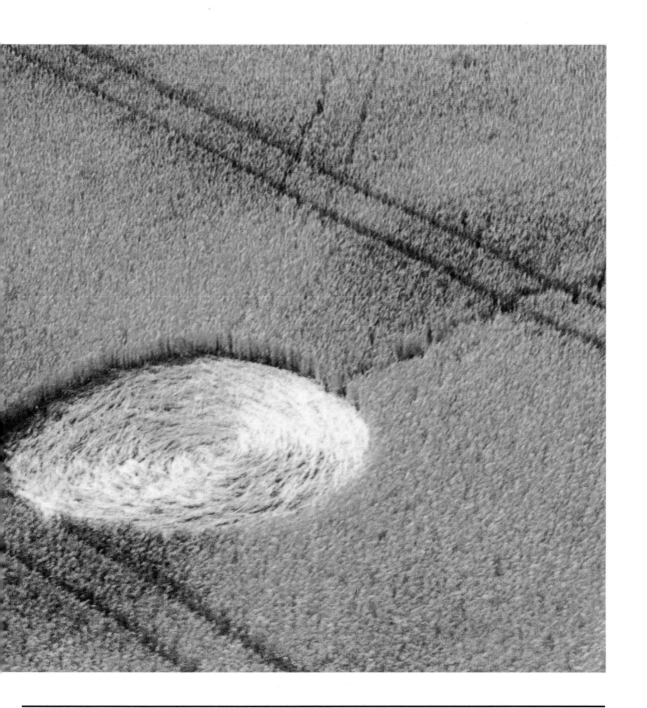

Plate 40.

Plate 41.

BECKHAMPTON, 1987

Beckhampton in Wiltshire is just down the road from the largest ancient man-made earth mound at Silbury. Ten circles of various dimensions appeared there in a field of wheat in August 1987. A very large set of five, 62 metres in diameter, was the first to be spotted. It had been created by a powerful force similar to that responsible for other discoveries in Wiltshire in previous weeks. Plants had been ripped out by the roots, with soil still attached, and thrown around the surrounding area. Close to the set of five was a small circle, 7·8 metres in diameter and swirled anti-clockwise.

About 250 metres to the south-west in the same field was a set of three circles in a straight line (see Plate 41). These were thought to be five weeks older than most of the others. The maximum diameter of one of the smaller circles was 7·56 metres. The floor was twisted around anti-clockwise into a tightly formed swirl pattern. The second of the smaller circles was 7·4 metres in diameter, but the similarly twisted floor pattern was swirled clockwise. The central circle was larger than the others, the diameter being 17·2 metres, and this time the plants lay in a tight clockwise pattern. The formation covered a distance of 35 metres in a straight line.

The area is surrounded by ancient

archaeological sites and has a very special historic atmosphere. 'These huge formations have never been seen on my land before,' said Mr Horton, the farmer, 'though the previous owner, Tom Wilde, saw the outline of ancient burial circles in this field during prolonged dry spells.' This is quite common in southern England, and is caused by the methods of building construction used by early man (see page 163). He continued, 'But these dramatic circles across my field are quite different. They have appeared within such a short period of time, not there one day but commanding attention the next. When I first saw them I thought some kind of whirlwind might have caused them, then I looked closer. It was different. I haven't a clue how they got there.'

David Burden from Chippenham was observing from a helicopter, with an experienced pilot at the controls, when they approached the Gallops just west of Beckhampton. David caught sight of this same stunning line of three circles. He told me, 'They were impressive. Neither of us could recollect ever seeing anything before to compare with them.'

When they went in to look more closely, they could see the well-defined edges to the tightly swirled Catherine-wheel circles. David went on to say, 'We were certain they could not have been caused by helicopters, as some people have suggested.' He was quite correct. We have investigated this theory thoroughly and it is totally discounted. I was fascinated by a comment David made to me later. He said, 'I felt just as though I had won the Pools when I first saw those indentations. It was a very special moment. I did not sleep that night, that was all I could think about.' 'We know the feeling,' I told him.

The helicopter rose up to about 180 metres and they then saw the other circles we had already investigated near by, and a new one almost under some high-voltage overhead electrical cables. David Burden gave us valuable information, because the field had been harvested by the time I interviewed him. He said, 'This circular area was not as fully pressed to the ground as all the others. It looked as though something had tried to land and then decided to go back up.'

WESTBURY,
1987

Captain D.F. Borrill of 658 Squadron, Army Air Corps, at Netheravon, was participating in large-scale military manoeuvres in his helicopter when he saw two circles in a field near the White Horse at Westbury on the north-west edge of the largest military ranges in Britain at Salisbury Plain. A few days later, on 7 August, he sent a memo to his senior officer to say that he had noticed a further three circles when flying over the area again. Plate 42 was taken by me minutes after I discovered these new additions on Tuesday, 4 August. Captain Borrill's helicopter can be seen photographing the mysterious indentations.

By now so many circles were appearing in southern England that we were having difficulty in visiting each new one immediately, as we liked to do. It was not until four days later that we managed to carry out our usual site investigation at Westbury. Something was different, there was another oddity. The southernmost of the three circles was swirled anti-clockwise when found, but was now confused. We checked the details against my photographs and those taken by the helicopter pilot when they had first been found. This proved clearly that a second event had taken place within this circle, while the other two had grown in diameter by 1·5 metres and both now swirled clockwise. The

two smaller circles were 7·75 metres and 9·8 metres in diameter. The largest was swirled round in a clockwise direction, while the southernmost circle now consisted of numerous twisted pinnacles and there was general confusion on its floor. It had been clearly anti-clockwise and neatly constructed when first found. The large central circle was just over 14 metres in diameter and was made up of plants swirled clockwise in a single revolution. Many plants within each area had been broken and thrown out on to the tops of surrounding plants. When first found, a 1 metre wall of standing plants had separated each circle, but now all three were merged.

Nine totally mysterious circles had now appeared in just five days in this area. Geoff Cooper farms opposite the field in which both sets of circles (three and two) had been found. He told us he had owned the field a few years ago before selling, and had been finding circles in it for the last seven years. He said, 'One night our dog went silly barking, real nasty like. Usually he would stop on a firm command, but not that night. He went on for ages, really upset he was. I wish now I had looked out because when I did in the morning I could see five circles had appeared in the corn during the night. I don't know what caused them. I don't think they are made by people. We have tried to make them with ropes, poles and so on, but they just cannot be replicated.' We agree with Mr Cooper. We have witnessed hoaxed circles and rings and studied all aspects related to human involvement at ground level (see pages 152–171) and have had to discount the hoax theory.

Plate 42.

WESTBURY,
1987

Plate 43.

The first three circles at Westbury (see pages 73–77) had been formed on the night of 3–4 August. The same night one was formed on the other side of the road in a field owned by Geoff Cooper. He had never seen one on this side of the road before.

My father was thrilled to be the first to find it. Plate 43 was taken at the moment it was discovered. People may wonder why we and our friends find so many of these circles. It is quite simple. We have researched the subject thoroughly enough to know what we are looking for. We spend a great deal of time specifically searching for them, so the odds are higher that we will find them.

What a splendid pattern this one had. Plate 44 shows the 7·7-metre-diameter, anti-clockwise swirl at the time of investigation. While the standing wall is unaffected, swirled spikes lean against it as they spiral out from the centre and strike the perimeter. Three similar circles had appeared in the field opposite in 1982, forming an equilateral triangle. Days later two more had been discovered below Cley Hill a few miles away. On those occasions several people had reported seeing a huge, circular UFO with numerous orange lights around the perimeter and spokes splayed out from the edges. It moved slowly across the sky between the two fields where the circles were found hours later. Although it is too soon to draw conclusions about a possible connection, reports strikingly similar to this have been received from many sources, which may begin to suggest a connection.

Plate 44.

Plate 45.

UPTON SCUDAMORE, 1987

Barry Dyke, a flying instructor, was flying towards Warminster on 6 August. A few miles before the village of Upton Scudamore he was amazed to see what looked like five dots on the ground. 'I reduced height and adjusted my course to fly directly over them,' he said. 'I could see they were just like the circles my colleagues had reported seeing across southern England.'

By chance Don, Busty and I met Mr Dyke at Old Sarum airfield the next day, and he told us of his discovery. We changed our flight plans to take us over Upton Scudamore. Plate 45 was taken as we approached from Warminster. Plate 46 was taken overhead and shows the minor road to the village at the bottom left. This flight was unforgettable. As well as these five, we found four more circles near Warminster. We would have liked to search further, but fuel was running low and we had to return to the airfield. We went straight home to collect our equipment and headed back to Wiltshire directly to the set of five. Plate 47 shows Pat, who had joined us, examining the vast number of plants thrust out of each circle on to the surrounding wheat. 1987 was the first year we found damage of this kind.

The four satellites were spaced in a square around the large, central circle. Each one was

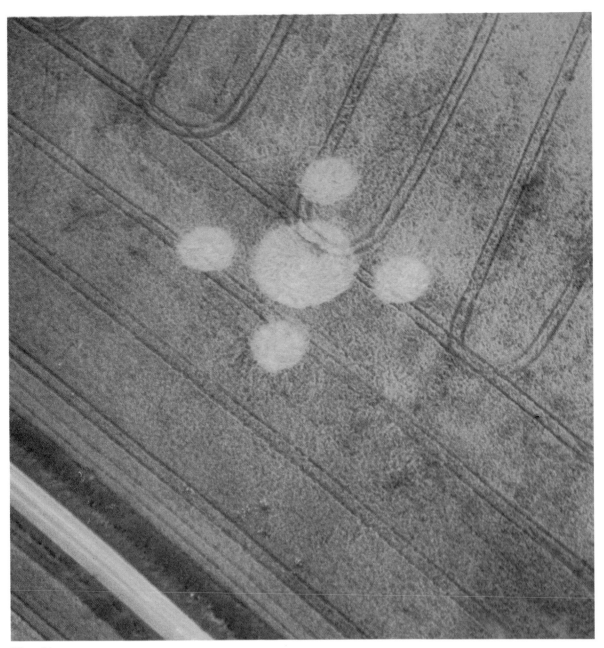

Plate 46.

about 7 metres in diameter, which was much larger than those found in similar formations before. And the floor patterns were different: 'S' shaped radial swirls licked out from each centre. In one satellite the head of the 'S' lay clockwise at the perimeter, in the other three the 'S' was reversed, with an anti-clockwise swirl towards the edges. The central circle was 12·74 metres in diameter and also had a reverse-'S' floor pattern.

Frank Brake had found a circular area in a field of yellow charlock near Upton Scudamore in 1974, before our research began. On closer inspection, he found the area had been mysteriously consumed overnight. The plants had gone and the bare earth had been visibly disturbed by something extraordinary from the air. The day before the plants had been 1·3 metres tall and in full flower.

A high number of UFO sightings are reported around this village, some of which are extraordinary and witnessed by groups of people.

Plate 47.

Plate 48.

WESTBURY, 1987

This circle at Town Farm, 15.6 metres in diameter, had one of the most spectacular floor patterns we had seen. The radial swirl was so impressive that it seemed too good to walk on. It was like the effect created when an expensive rocket explodes at high altitude on Bonfire Night, throwing out hundreds of radial lines of burning phosphorus. The circle contained half a revolution anti-clockwise swirl, as though whatever had produced it was revolving. Plate 48 shows Pat and Busty inside it with the White Horse on the hillside in the background. Some of the wheat stems were broken, and many had been thrown on to the surrounding area. The force involved was so great it had caused the dry, crusted, surface soil to break up into powder.

We gathered at Old Sarum airfield and planned a flight which would also take us over the five circles at Upton Scudamore which had been formed in the night of 6–7 August. Plate 49 is an aerial photograph showing the field which contains the circle shown in Plate 48. It also shows many other fields in which circles have appeared in recent years. The most recent circle can be seen bottom centre, two days after it appeared. Our trip was memorable. Not only did we observe and photograph these six circles, but we found others. We were confident that fifteen circles had appeared around Warminster in less than twenty-four hours. Most were less than twelve hours old when discovered.

I was filing my photographs of this circle when I noticed an unusual marking on one of them, some 35 metres south-west of the circle. Under the magnifying glass it looked like a narrow, single ring. I phoned a colleague who lived close to the site, to ask him to check it out. Two hours later he called to confirm that it was only the second known independent ring yet found, and like the other one at Kimpton, near Andover, it was in a field with a circle near by. The same phenomenon appeared to be responsible for both circles and rings.

We have investigated thoroughly a photograph taken by Busty of Pat and me in the circle, in which a white disc-shaped object can be seen about 18 centimetres from the swirled centre (see Plate 50). Pat and I both seem to be looking at it, though neither of us can recall why we were looking that way. We saw nothing like it at the time. The sun had just set behind Busty when he took the picture, which would seem to rule out light refraction. We have taken highly magnified prints of this object, which also appears on the negatives. They show a clear, disc-shaped object, bright white in colour and with well-defined edges. We have had the negative analysed professionally, but no logical explanation has been found. It is another facet of the subject which we are unable to account for. We have learnt the importance of flexibility.

Plate 50.

Plate 49.

Plate 51.

BRATTON,
1987

You can see in Plate 51 the first double-ringed circle we found. The team are investigating inside the circle and both rings can be seen clearly, with the hill rising up behind to Bratton Castle. This huge formation, 30·28 metres in diameter, was one of many to appear in this area on 8 August. The large circle, 16·3 metres in diameter, consisted of flattened wheat plants swirled around in a tight two-and-a-quarter-revolution Catherine-wheel swathe. The plants were not as firmly pressed to the ground as many we had seen before. The inner ring was 1 metre wide, though this varied by a few centimetres in places. The wheat had been swirled anti-clockwise. The outer ring was slightly wider, but here the plants were pushed down clockwise. The standing walls surrounding the formation were as upright as ever, apparently unaffected by the event.

We checked data from the British Meteorological Office for the possible time of formation and again drew a blank. The wind direction varied from 280° to 270° and the mean speed from 6·3 to 9·3 knots. 8 August was overcast with only 0.1 hours of sunshine and a maximum temperature of 16·9°C. An extremely small amount of rain had fallen. There is a high hill near by to the south of this field, which ruled out formation by some new and unknown weather vortices forming in the lee of the hill.

The Wingfield family were enjoying a day out in the beautiful countryside when they spotted the circle and our team below the Bratton escarpment, on which they were standing. George Wingfield, his wife and two young sons walked down the steep hillside towards the edge of the field. What happened next frightened Mrs Wingfield so much that she did not want to go any further. Her family were some yards below her and, as she lifted her eyes from the steep bank, she was stunned by a blue flashing light coming from a point just above the ground in front of her. The light was pulsating as if it was reflecting off a spinning shiny surface.

Mrs Wingfield gave us a written statement afterwards. 'I had suddenly felt uneasy, and as I walked down I saw blue flashes on the ground occurring regularly. This happened with the evening sun shining and was quite distinctive, but was not seen by my husband or sons, who were further down the hill. The flashes appeared to be a reflection from something unseen and occurred every second or possibly with shorter frequency. They were like the reflections from a blue reflector on the slowly rotating blades of a helicopter. When the sun passed behind a cloud the blue flashes ceased.'

Plate 52.

CHILCOMB, 1987

Six days after the circle and two rings appeared at Bratton, another most impressive circle with one ring was found at Chilcomb near Winchester. As can be seen in Plate 52, the floor pattern in the circle was notably different to the tightly swirled floor at Bratton. It gave the impression of an enormous vertical pressure downwards which fanned the plants radially outwards with tremendous force on impact. As usual, the standing walls around remained untouched. In these circumstances you would expect to find degradation, but there never is any. The diameter of the circle varied between 19·45 metres and 19·11 metres, which is common. However, the floor pattern was not; it was radial and the only hint of direction was clockwise and towards the perimeter. The inner band of standing plants varied in width from 1·8 metres in the south-east sector to 2·26 metres in the north. The ring was 1·64 metres wide in the south-east but as little as 1·26 metres in the north-east and it flowed anti-clockwise. A number of plants were lying down westwards out of the inner standing ring in a trumpet shape, the wide end of which ran into the flattened ring and appeared to have been formed at the same time. We traced the narrowing band of plants back into the standing band and found that it ended at a point where only one plant was involved. We had seen a form similar to this in a single ring near Kimpton (see pages 62–66). What part this had played in the original formation remains a mystery.

When Pat and I first arrived in the field a fallow deer leapt away from the edge of the circle and through the field to the safety of woodland below Telegraph Hill, overlooking Winchester, the ancient capital of Wessex. This was one of the pleasures of site work.

Another unexplained peculiarity arose out of our research at Chilcomb. I was looking at photographs taken by Busty in the circle on 16 August. On one of the prints were what looked like two black ribbon darts, similar to the head of the forked spear usually associated with the devil. One was clearly focused, but the second was blurred as if by rapid movement. They appeared suspended on the same axis above the circle, at roughly 25°. Experts we have consulted have been unable to explain their appearance on the negatives, and have ruled out foreign objects in the camera or on the lens. Technical explanations suggest that the objects were recorded on the film so were in the focal field of the camera, that is they appeared physically over the circle close to the researchers. By now most of us had experienced strange noises or seen blue flashes, but we had never seen anything like this. We believe it is important to record all the evidence, and not ignore the more bizarre occurrences simply because they cannot be explained rationally. By remaining open-minded we may eventually be able to explain the 'circular evidence' which is appearing throughout the world.

The mystery deepened when George Wingfield visited me on 10 September. It was Mrs Wingfield who had seen blue flashes near a similar formation at Bratton (see pages 94–95). I was showing him some photographs, including this one. 'There are marks like that on a picture I took inside the circle at Bratton,' he said, 'in the same place on the print as here. My son asked what they were and I told him they must be dirt on the film.' He went home, promising to find the print. He phoned a few days later, highly embarrassed. 'I don't have a print with those marks on. I'm usually very methodical and have a good memory, but I've checked all my prints and negatives and they don't show the black darts. I was so disturbed by this that I checked with my son. He didn't know what I was talking about. I'm beginning to doubt my sanity and can only conclude I dreamt this.'

All this was strange, but it does not end there. When Pat first saw the photograph, he also said, 'I've seen this before. I think they are on a photograph of mine.' He too has checked, but cannot now find the photograph.

At this point we should mention the famous encounter reported by Joyce Bowles and Ted Pratt in 1976 when a UFO was said to have landed near the Chilcomb lane and a humanoid approached Mrs Bowles's car. It was wearing a silvery suit and had pink eyes. The car engine had cut out in the lane just 1,500 metres from the field we were now visiting. 1976 saw the first circular areas being discovered around Winchester, and the 1976 circle was discovered just down the road from where the UFO was seen.

BRATTON,
1987

Plate 53 was taken from the air at 4 pm on 22 August. The set of five circles at the bottom had appeared the day before, adding to the thirteen already in the fields below the escarpment. Others can be seen in the adjoining field. The combine was just finishing harvesting the field in which the latest set had mysteriously appeared. By the time the plane had returned to Old Sarum airfield, the field had been harvested, removing all virgin evidence. Dr Meaden of TORRO visited the site the following day and found the affected areas still clear enough to measure, so all was not lost.

The stems of the wheat had been pressed so firmly to the ground that the combine had not been able to cut the plants involved. The large circle was 15·2 metres in diameter with an anti-clockwise swirled floor. Satellites north and south of this circle were 6·5 metres in diameter, one swirled clockwise and one anti-clockwise. The satellites to west and east were just over 7 metres in diameter, and again one was clockwise and one anti-clockwise. The distances between the satellites and the central circle varied: they were 10·9 metres (north), 10·95 metres (west), 11·1 metres (south) and 11·9 metres (east). The diameter of the formation was 50·63 metres. Dr Meaden kindly provided these details. We should say here that we do not share his belief that there is a meteorological solution to the problem. The evidence does not seem to support the theory.

Plate 53.▷

Plate 54.

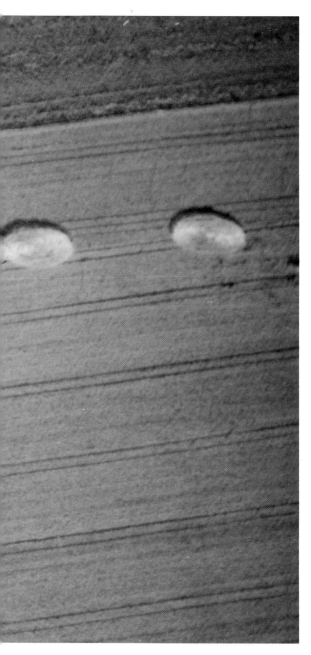

WINTERBOURNE STOKE, 1987

Having studied circles for some years, we now know the areas most affected, so, when planning flights, we cover as many known sites as possible in one run. While on a route south-west of Stonehenge flying towards Warminster, we were astonished to see below us a huge circle and ring with a further three circles near by. We had received no reports from this site before. Plate 54 shows the marks as we found them at 3.30 pm on Saturday, 22 August 1987. They were truly spectacular. Two hours later we reached the field by car.

We went straight to the enormous centre-piece, a circle 25·3 metres in diameter in a field of rye. All the ripe yellow plants were compressed hard into the ground, forming a three-and-three-quarter-revolution clockwise swathe. Extremely heavy pressure appeared to have been involved, although no damage was apparent. The cut off between the floor and the surrounding crop was abrupt. The ring had a symmetry we had not seen in many other rings. The plants lay swirled anti-clockwise in the 1·3-metre-wide band and the lay was tangential in a typical centrifugal fashion. It looked so real that you could visualise it spinning. There was a toothed bay, 28 centimetres deep, cut into the standing wall in the southern sector. The angle at which the plants were lying on the floor changed on a line from the bay to the centre of the large circle. The ring of standing plants between the circle and the flattened ring was 2·3 metres wide.

To the west of the ring and circle was the closest of the single three circles (see Plate 55). It was a particularly beautiful clockwise swirled circle, 6 metres in diameter. The other two circles to the north were larger, with 8-metre diameters. One swirled clockwise and the other anti-clockwise. We estimated they had been formed the day before. Their site was one of many overlooked by a busy road – the A303 from London to the West Country passes this field on a high embankment, allowing a view of the field concerned. Nothing unusual was reported until eight weeks later.

Two months later, on Thursday, 22 October, a Harrier GR5 jump jet, piloted by Taylor Humphrey Scott, took off from Dunsfold in Surrey for a test flight over Salisbury Plain. When it reached this spot at 5.06 pm, west of the top-secret aircraft establishment at Boscombe Down, radio contact was suddenly lost. The pilot had just passed a routine message to the control tower at Boscombe Down which did not indicate any problem. All aircraft in the area were alerted and finally an American transporter sighted the Harrier ninety miles south-west of Ireland. The crew were astonished to see that the aircraft was pilotless and

the cockpit canopy was missing. They shadowed the jump jet for 410 miles before it eventually ditched in the Atlantic 500 miles out to sea. A huge air and sea search was set in motion to find the missing pilot. His body was found on the evening of 23 October in a field overlooking the mysterious circles at Winterbourne Stoke. His secondary parachute was near by and a dinghy from the aircraft was in the corner of the very field in which the circles had equally mysteriously appeared out of the blue. The aircraft had changed course by several degrees when the incident took place over this field. Air experts reported they had no idea why the experienced British Aerospace pilot should have ejected.

The following are the known facts. Four circular areas of flattened corn appeared in this field on Friday, 21 August 1987; most of our evidence tends to indicate a mysterious aerial component is responsible. On Thursday, 22 October £13·5 million of the very latest aerial technology, a Harrier jump jet, mysteriously loses its pilot over this same spot. His ejector seat has never been found and is believed to have remained in the aircraft.

With the Ministry of Defence, we are left to ponder two inexplicable events over the same field within weeks. Their evidence lies below the Atlantic Ocean, ours is pressed firmly into the field concerned.

Plate 55.

Plate 56.

CHEESEFOOT HEAD, 1988

At about 8 am on Saturday, 4 June I reached the Punch Bowl to check for new circles. I had checked this site many times in the previous weeks, so was amazed to see a pattern of three. My first view of them was a glimpse through the trees as I drove along the road bordering the amphitheatre. This first sighting of new circles there is always tantalising until you pass the trees and the whole field comes into view, with the pattern of the group before you. It never fails to thrill me. I parked and walked across the grass verge to study what lay before me in the wheat crop. The three circles, in a triangular formation (see Plate 56), were situated roughly in the middle of the field below me. This year the farmer would not let anyone enter his fields to study the circles, so I had to assess their diameters and centre distances from outside. I took photographs with a zoom lens.

Through the binoculars I first ascertained that the rotations of the circles were all flattened clockwise. To find the diameters I walked around the top of the hill to look along some faint tractor lines. I knew from the farmer that the lines in this field were 20 metres apart, and by marking their spacing on a piece of card 0·5 metres from my eye I had a scale to find their diameters, which proved to be 10 metres. This triangle was a new phenomenon.

Plate 57.

CORHAMPTON, 1988

Little did I know when I answered the phone on 8 June that this was the first of a staggering number of circles to be reported during the next two months – over one hundred. Charles Hall had previously found single circles twice on his farm, but he was quite shocked when he spotted three large ones in a field of winter barley.

I left my office in Andover and was soon with Mr Hall at the field. There were three circles, each 10·8 metres in diameter. About 35 per cent of the barley plants were compressed to the ground and swirled clockwise. The pressure swathes from the centre of the circle expanded towards the perimeter. The remaining plants had clearly also been flattened when the event occurred but had now lifted back to the light in an extraordinary fashion. Later when we saw the circles from the air we found that they were at the points of an equilateral triangle and the circle edges were separated by 1·9 metres of standing crop. The compass bearing through the centres of two of the circles was precisely magnetic north.

The plants had grouped themselves into oblong areas about 60 by 30 centimetres, and in three groupings the plant stems, along which there are three nodes or knuckles, were lifting up in a selective manner. In the first group the plants were bending up on the node nearest to the roots to seek light. In the second group plants were lifting up on the node half-way up the stem, and plants in the third group were bending on the node close to the head of the plant. These groupings were repeated throughout the affected areas in each circle. I wanted to see this from the air, so I arranged a flight with Busty from Thruxton an hour or two later.

We have experienced one surprise after another during our research, but few have matched the shock we had when we looked down on the Corhampton triple formation for the first time (see Plates 57 and 58). The plants had now grown into a pattern consisting of seven concentric rings and forty-eight spokes. Each circle had the same form. We later found that plants inside an identical formation a few miles away had formed the same dartboard or cartwheel pattern.

I discussed this at length with Dr Mark Glover, who is a plant and agricultural specialist. We visited the site and, after studying his findings, confirmed those of other laboratory tests we had carried out. 'I cannot find any explanation for those particular crop areas which had been down on the ground for at least a week but no more than two weeks,' he reported. 'I can, however, say they are not caused by any of the following: cultivation, fertiliser applications, agrochemical treatment, disease, pest damage, soil type or plant growth habit. I find the problem fascinating and look forward to seeing any further discoveries.'

Plate 58.

Plate 59.

CHEESEFOOT HEAD, 1988

On Saturday, 25 June I was delighted to find a large circle with two rings round it (see Plate 59). This was not a new formation, but it was the first time it had appeared in the Punch Bowl. It was three weeks since I had found the triangle of circles on 4 June and the crop was now turning golden, which enhanced the marvellous precision and sharpness of the double ring.

The new formation was much closer to the road than the triangle, so I could see easily that the circle was flattened clockwise, the inner ring anti-clockwise and the outer ring clockwise again. The circle was 23 metres in diameter, the two standing bands 2 metres wide, the inner ring 1·3 metres wide and the outer ring 1·3 metres wide. The two groups were very photogenic and I saw a number of people taking photographs of them at different times.

That morning several military vehicles, which were being used in local manoeuvres, were parked near by. I told two officers about the circles and suggested they look at them as they were so spectacular. One of them, a Major, said he would be able to fly over them that afternoon, so I asked him if he would let me have copies of any photographs. He agreed and I gave him my phone number. That afternoon my daughter and I visited the Punch Bowl again. As we left the car, an army helicopter flew low over the field, circling it twice. The occupants were obviously very interested and taking photographs.

I was pleased the Major had taken up my suggestion and looked forward to receiving some good photographs. However, I heard nothing. I later enquired at the local helicopter base, through a friendly and previously helpful military acquaintance, but to no avail. Nobody knew of the Major, whose name I supplied, of the flight or of any photographs of the Punch Bowl.

Plate 60. Plate 61.

SILBURY HILL, 1988

In only eight weeks an incredible fifty-one circles were found within seven miles of this famous, ancient, man-made earth mound, the largest in Europe. Fifteen of them appeared within a few days in a wheat field opposite the mound and close to the A4. The first five, which were photographed from the mound, appeared during the night of 14–15 July.

Twenty-four hours earlier Mary Freeman, who lives in Marlborough, Wiltshire, was leaving the ancient stone circle at Avebury not far away. At 11.13 pm, with about 70 per cent cloud cover, she saw through the driver's window a large, golden, disc-shaped object within the cloud. A bright white parallel beam of light came from the bottom of the disc at an angle of roughly 65° and shone across the sky towards Silbury Hill, one and a half miles to the south. 'I was amazed, but not frightened,' she said later. 'It was ethereal.' Suddenly about half a dozen articles, which were in a recessed pocket along the dashboard, rose up and flew backwards, landing in her lap and on the passenger seat next to her. She said, 'It was as if a surge of energy passed through the car.' Her report is similar to a number of others reported in Great Britain above circle fields.

Plates 60 and 61 were taken in the wheat field a few days later. Two sets of five had now appeared and a further five single circles were formed over the next day or two. The tracks visible between the circles were made by us while taking measurements.

The first set of five consisted of a large clockwise-swirled circle, 17·2 metres in diameter, with satellites about 6·5 metres in diameter. The satellite at 21·5° swirled clockwise, whereas the other three, at 112·5°, 202° and 244°, swirled anti-clockwise. The formation was 88·15 metres across. An 11,000 volt overhead electric cable passed over the set of circles. We found out from the Electricity Board that there had not been any voltage surges or loss of supply that night.

At noon on 10 September the phone rang at my home. 'Hello, it's Geoff Smith from Charity Down Farm, Goodworth Clatford. We've found a huge circle with a very large ring round it, next to the one which appeared the other week. There are four small circles at the compass points around the central circle, but they're positioned centrally round the ring. It's really very strange. We're cutting the field now. Can you come over? There's something else here that's very odd . . .'

And so it goes on. We know much more now than when we started, but much remains to be evaluated.

CHAPTER TWO

———◇———

CIRCLE FORMATION

The formation of circle groups has probably occurred for hundreds of years. Unfortunately, we have no written or pictorial evidence. There are no detailed descriptions or photographs of the circles that are reported to have appeared from the 1950s right up to the 1970s. The only information we have about flattened circles in crops many years ago comes from farmers and people living in cereal crop areas. About seven farmers have confirmed that single circles have appeared in their fields over the last twenty-five years; a few report them from as early as 1940.

Figure 1 shows our investigations so far.

A: the single circle appears with a wide range of diameters, from 3 to 22 metres. This basic single circle is formed with either a clockwise or counter-clockwise swirl. The tightness of the spiral-swirl whorls varies from 4 turns to straight radial swathes.

B: the unequal double. The diameter range of the small circle is 5 to 7 metres, and of the large 12 to 14 metres. Spiral turns range from 1·5 to 3 times.

C: the triple. The large circle is always in the centre, and has a diameter range of 16

Figure 1

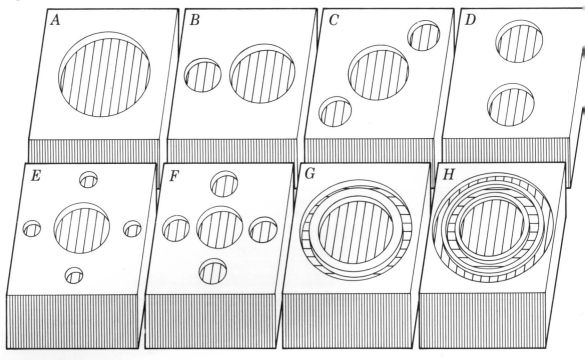

to 18 metres. The diameters of the smaller circles range from 8 to 10 metres. These circles have all had clockwise swirls, or 2 circles have clockwise and one of the outer circles has an opposite swirl.

D: the equal double circle is quite rare. Those investigated have been small, around 4 metres in diameter.

E: the classic quintuplet. The large circle varies from 17 to 20 metres in diameter. The equidistant satellites are 3 to 4 metres in diameter. All the circles are usually swirled clockwise.

F: the large satellite quintuplet. The large circle varies from 13 to 14 metres in diameter. The satellites' diameters are 7 to 8 metres. Some satellite circles have contained a very different rotation swirl to others in the same set. Each satellite may vary considerably in the tightness of the swirl.

G: the beautiful ringed circle. This may contain either type of swirl, or mixtures of both or multi-layers, and is usually compressed

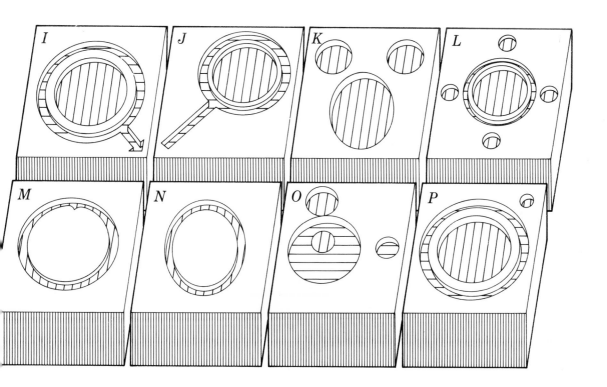

very hard to the ground. The diameter of the circle ranges from 17 to 20 metres. The width of the standing band of crop varies from 1 to 2 metres. The ring may be swirled either clockwise or counter-clockwise and varies in width from 1·3 to 2 metres.

H: the remarkable double-ringed circle, of which only one example is known. The circle was 16·2 metres in diameter and swirled clockwise. The first standing band averaged 2·3 metres in width. The first ring width averaged 1·1 metres and swirled counter-clockwise. The second standing band averaged 2·3 metres in width, and the second ring averaged 1·2 metres in width, swirled clockwise.

I: ringed circle, male symbol, of which only one example is known. This ringed circle was of the type stretched into an ellipse (see page 134). A pathway extended away from the ring, which terminated in an arrowhead shape. The circle diameter averaged 19 metres, was swirled clockwise but also possessed an outer band of counter-clockwise swirled crop, averaging 2 metres in width. The standing band averaged 1·8 metres in width, and was wider at the stretched side. The ring averaged 1·5 metres in width, was wider at the stretched side and swirled counter-clockwise. The pathway was flattened away from the circle centre.

J: ringed circle with elongated path. The elongated path of this type can be either straight or curved, and we have investi-

gated examples of both. The circle diameter ranges from 18 to 20 metres. The standing-band width varies from 1 to 2 metres. The ring width is from 1·5 to 2 metres. This type of circle has, so far, always contained a clockwise swirl. The ring is swirled anti-clockwise. The pathway ends abruptly and its length ranges from 20 to 30 metres.

K: a large circle with twin satellites positioned on one side. Only one example is known. The large circle was really an ellipse with a major axis of 21·5 metres, and a minor axis of 15 metres. The swirl was as type G. The two satellite circles were both swirled counter-clockwise.

L: ringed circle quintuplet. Again we know of only one example. The circle was approximately 14 metres in diameter. This approximation is due to the effect of some wind damage. The standing-band width was 1·5 metres. The ring width was 1·5 metres and it was swirled counter-clockwise. The four satellite circles averaged 3·5 metres in diameter and were all swirled clockwise.

M: single ring, of which only one example is known. The diameter averaged 10 metres. The ring was slightly egg-shaped and one half swirled clockwise, the other half counter-clockwise. This contra-rotation commenced at the heart shape at the top of the circle, see Fig. 1 for clarification of the swirl. The width of the ring averaged 0·5 metres.

N: an elliptical ring. One example is known.

The major axis was 7·6 metres, and the minor axis 6·2 metres. It was swirled clockwise.

O: superimposed circles. This configuration consists of two B-type formations, one superimposed on the other. The dates of the two discoveries were two weeks apart. Confusion of lays takes place when the later, superimposing circle tries to re-lay the existing flattened stems.

P: a formation that receives an additional circle or disturbance. The addition was another circle, in one case, and a distorted circular shape in two others. There may have been other instances in previous years when these additional opposing swirls occurred. We were not aware, at the time, that a further addition had taken place.

Superimposition presents a further problem in circle and ring appearances. Some triple and quintuple groups of circles have been swathed in a confusing manner, suggesting a second forceful visit.

Sometimes, when we visit a group of circles one will have a predominantly clockwise or counter-clockwise swirl. About three-quarters of it seems to have been subjected to a force attempting to brush the swirl in the opposite direction. The upper half of the affected stems has been bent violently backwards or whole bundles have been partly swung around to face a different direction.

In some instances, this re-laying may have occurred directly afterwards as part of the crop's initial flattening. It looks like an addi-tional pass of the force, only affecting one circle in the group. In other instances we know some circles have been re-affected later because the stems have a completely different lay to the lay recorded on our first visit. Despite this double exposure, the crop usually continues to ripen normally and the heads remain intact.

It has been possible to record these detailed descriptions because the evidence remains on the ground, in some cases for several months. Even after harvesting, the circle floor can be seen quite clearly because the gathering level of the combine harvester has not been set low enough, allowing much of the lay to be seen in its original form. Sometimes ripened seed has fallen to the ground. During November the seed will germinate, exhibiting a green circle of young plants in the ploughed ground. This is often oversown and disappears with the new crop growth.

Stubble burning after the harvest preserves the remains of circles too. Providing rain has fallen before the burning, the compressed circle floor will remain damp enough to prevent combustion. The remains of the circle floor can be seen amongst the blackened surroundings of charred stubble.

CIRCLE FLOOR PATTERNS AND LAYS

If we had found only one type of swirled circle, it would probably be much simpler to theorise about it. For instance, if there were only single circles always swirled clockwise we might be able to understand their creation by applying known scientific criteria. If the manner in which the circles are created remained constant, various theories could be eliminated.

Before the late 1970s it looked as though single circles were all we had to consider; but, as has always been the pattern, and as we have learnt over the years, something, maybe some intelligent level, keeps one or more jumps ahead of what we think are answers to the various plant-stem details we have noted. The problem of finding answers is multiplying each year. Practically every new site holds a fresh detail. It is little wonder that we find trying to fathom out the phenomenon an enormous task.

In a circle, when you are confronted with a plain clockwise flattened swirl, you can imagine that a spinning vortex could have created it. The force field of the rotating column would have to pull or push all the stems of the crop down together. The movement from vertical to horizontal has to be gentle, so the stems are not broken by a whiplash effect, but it has to be very strong to lay the stems in such a hard compacted manner without crushing or damage. The stems are laid horizontally immediately above the soil surface. Sometimes the whole root system of each plant is slightly tilted over in the direction of the lay.

Considering the above facts you may be able to produce an adequate theory. This is fine until you discover a flattened swirled circle in an oil-seed-rape crop. The stem bases of these plants are between 8 and 16 millimetres in diameter, and are bent over at the base to almost 90° without breaking. If you try to bend one of them, it snaps like a carrot. Attempt to straighten out a bent stem and, again, it will snap.

It may seem to be feasible that there is an earth or atmospheric force available to create a swirled circle. Fine, but the swirl is not always clockwise. Sometimes circles have a counter-clockwise swirl, sometimes a mixture of both, sometimes no swirl but straight radial flattening from the centre of the periphery, as the spokes of a wheel. There are almost as many patterns of swirls, or non-swirls, as there are circles; the permutations are considerable.

Before considering these 'lays', a description of whorls, or veins, is appropriate. When the lay of the stems is greater than tangential, that is, the heads of the plants are further away from the centre than their roots, a characteristic whorl or vein is formed, which is a long bundle of gently twisted stems, similar in construction to a rope. The twist of the whorl, followed from the centre, is the same rotational direction as the lay of the floor pattern. For example, if a circle has a clockwise swirl, its whorls will be twisted clockwise. Starting from the centre, the twist of the whorls grows less as they extend out to the wall. The opposite occurs in a counter-clockwise-swirled circle floor.

The plates and their descriptions on pages 136–145 give many examples of whorls and spirals. The number of turns or coils varies with different circles, the maximum being about five and, of course, this number decreases until the straight radial pattern is reached. The straight radial swathe contains no twisted whorls.

The twist rate of whorls is not constant in circles containing the same number of coils. One four-coil pattern may have a high twist rate, while another has a low twist rate. The twist rate can be approximately one turn per metre in some circles, in others it can be anything less than this, and only disappears altogether in a radially laid circle. The whorl is also described as a swathe, but, whichever word is used, it is created in a spiral form. The whorl thickness or diameter is approximately 12 centimetres. The whorl is diffused into the general lay of the floor, so it is apparent but not always clearly defined. There is no definite edge to it from which to measure. The number of whorls in a circle cannot always be clearly counted. They start from the centre, but often break up into branches. Those that do not break up can be clearly counted. Divisions of a whorl can occur several times before it reaches the wall.

The heads of plants are rarely laid closer to the centre of the circle than their roots, although exceptions to this have been observed

on a few occasions and just how the plant was laid with an inward lay is a puzzle.

Only the lay of the floor and whorls have been mentioned so far, but already the considerable number of complications and varieties of floor patterns that have been encountered must be clear.

Figure 2 shows some of the many lays we have discovered over the years. The dot in each case represents the central area of the circle, the arrows indicate the direction of the plant heads away from their roots. It is difficult to understand what kind of force is capable of manipulating the crop stems into such configurations.

Figure 2

A: multi-turn clockwise swirl. This is the 'original' classic swirl observed from 1981 and thought to be, at that time, the set pattern of circles. The plants are usually compressed very flat to the ground and the whorls are significant. So far, the number of turns range from two to a maximum of six.

B: single-turn clockwise swirl. This classic pattern usually has a well-defined central swirl with well-defined whorls.

C: part clockwise swirl. This starts with a loosely defined central area. The whorls are not sharp but have a 'spread' appearance on reaching the wall. Here the swathes slightly penetrate the circumference wall, giving an uneven appearance.

D: straight radial. This pattern starts from the central area and continues to the wall in a straight line, with practically no whorling effect but flat swathes up to 0·5 metres wide. The lay is thinner between the swathes and emphasises the radiation pattern. The outer ends of the swathes tend to give the impression they have been forcibly thrown into the circle wall. This penetration has been measured up to 0·75 metres. The end of the swathe curves upwards. The strength of the force that throws these swathes into the circle wall is further emphasised by the small number of part and whole plants that are found hanging in the top of standing crop close to the circle wall.

E: a radial swathe with its outer end overlying a swirl band. A most curious combination, showing the result of multi-directional force fields. There are X and Y types. In the X type the radial swathe is similar to D, but its length falls short of the circle perimeter by about 30 to 40 centimetres. The space between the swathe end and the circle wall contains a swirled band of the crop, usually in the clockwise direction. The Y type again is similar to D. In this case, it reaches the circle wall, where there is a very thin line of swirled stems in the clockwise direction. This thin line consists of about 15 stems per metre of circumference and is not always a continuous ring.

F: this has the same characteristics as C but in the opposite direction.

G: single-turn counter-clockwise swirl. The characteristics are similar to B, but in the opposite direction.

H: multi-turn counter-clockwise swirl. All the characteristics are similar to A.

I: swirl-ended straight swathe. This is another curious lay. The swathe starts from the central area, as in D, but about 1 metre from the wall it becomes a swirl, complete with short whorls and very similar to the outer ends of whorls in lays A and B. This interesting flicked-end swathe has a streamlined appearance.

J: swirled start radial. This commences at the central area similar to B. The swirl then straightens out to become a radial and ends at or beyond the circle wall, similar to D.

K: the 'S' swirl. This is one of the most recent swirls to be recorded. From the central area of the circle the whorls start off in a counter-clockwise direction as in B. They then swing in the opposite direction in a most spectacular way, just like somebody combing their hair to create a neat wave.

L: offset radial. These swathes start at the central area in a tangential manner, then continue in a straight line to the circle wall, which they meet at an angle. There is sometimes very slight penetration of the wall.

M: contra-rotated swirls. This is the most astounding lay of any so far. Most of the circle floor is covered by a multi-turn clockwise-swirled lay. An outer 2-metres wide rim is swirled counter-clockwise. There is a small amount of overlapping at the junction of the contra-flowing stems. There are no standing stems on this line or gradation. The contra-rotated band has a swirl and whorls exactly as though it was the outer part of a full counter-clockwise swirl.

Circles with a serrated wall caused by swathe penetration have a further characteristic. Where crop stems are laid between penetrations, the stems tend to lean against the wall and splay to left or right in a fan-like shape. As the edge of the lay tends to die out along this arc, the wall does not have a sharp, upright appearance.

Some circles have multi-patterns; two or three of these are shown in Plates 62–6. These circles clearly demonstrate what a complicated laying process must have taken place.

One of our major problems is trying to understand close-proximity contra-rotation, that is a circle with a contra-rotated flattened swirl and no gradation between the opposing lays. This seems unnatural and, on the face of it, a deliberate attempt to display intelligent manipulation. A close inspection of the lay changeover line reveals a sharp demarcation between clockwise and counter-clockwise-laid stems. One stem is lying in one direction, its immediate neighbour, towards the outer edge, is lying the opposite way.

When plant stems are laid flat, they remain straight, and do not curve to follow the arc of their particular radius. The exception to this is where a tightly swirled central area is formed. This will contain some slight buckling of the stems. If the stems are laid in such a way as to form a pattern, similar to the grooves on a record, they could be described as tangential. Some circles will contain this tangential lay, while others will have their own angle of lay from tangential to a straight-out radial swathe.

We have discovered circles that contain multi-layered floors: up to three layers have been exposed by lifting the upper layers. The direction of this layering is often quite dramatic. For instance, we have lifted a top layer which is rotated clockwise and underneath we have found a layer facing in the opposite direction. Under this is a further layer which is radial. In spite of all this, no damage has been inflicted on these multi-layer, multi-directional flattened crops. To further complicate matters, only a small quadrant of the floor may be affected in this way, the rest being a conventional single swirl. Of course, the under layers of any circle are affected by lack of sunlight and sometimes become the victims of dampness and insufficient light. The density of the crop will effect this too, because a sparsely sown crop, when flattened, has a much better chance of drying out after a shower of rain.

Some circles have contained an additional enigma: a flattened 'path' underlying the main swirl, approximately 0·3 metres wide and radial. None of the stems are damaged, as is usual, but half of the path, from the centre, is facing outwards and the other half, from the outside, is facing inwards. The halves overlap in the middle of the path length.

We now have contra-rotation and opposing lays. In Figure 2, page 123, lay E is fascinating. This particular configuration occurred in only half a circle. The lay here was mainly radial until it was about 0·5 metres from the wall of the circle. Here it met a clockwise swirl at right angles to it. Some of the clockwise swirl was overlaid by the radial, so it must have gone down before the radial path was depressed.

1987 produced so many different lay patterns compared to previous years that we began to wonder if we had been inspecting circle floors thoroughly enough in the past. Looking through earlier photographs and recalling just how complex the previous lays had been, we are fairly certain the lay patterns in 1987 were far more complicated, in most cases, than we had ever seen before. Possibly this proves that it is an unnatural, non-repetitive, annual event. We realised early in the 1987 season that something

was different, not only in the lay patterns, but in the feeling we had on entering the circles. There appeared to be an atmosphere, hard to define.

The swirl lays of previous years had been orderly and depressed neatly, either clockwise or counter-clockwise. The only variation would be the number of turns of the spiral whorls. These have remained quite constant and are one of the fundamentals of the circle construction.

Lays are important because, by studying them, investigators realise they are looking at the results of a force that has no known scientific explanation. It must be remembered that, however complicated the floor lays of these circles, the crop roots, stems and heads are rarely damaged. Even when stems are laid in opposite directions there is no apparent buckling, so they must slide neatly past each other.

In some lays we have found the 'parting' or herring-bone pattern, although herring-bone sounds too geometrically balanced. In this feature, as can be seen in Plate 74 and Figure 4 (opposite), the stems on the side of a parting are lying almost parallel to the parting line, while on the other side the stems are facing almost 90° away from it. The force that created it must have extremely sharp polarising division to leave no gradations. In circles that have these partings, which vary in length from 1 to 9 metres, there are also swirls which could be any one of those shown in Figure 2.

The G lay in Figure 2 (page 123) has only been seen in 1987. The circle phenomenon is complicated enough without trying to figure out the force that can produce an S swirl, with its own variations from an even sine-wave curve of both half-loops to a large loop starting from the centre and finishing with a small loop at the wall. The end of the small loop also varies from lying in a curve, running around the circle wall, to a lesser curve that just juts into the wall. S swirls are also made up of whorls, the ends of which jut into the circle wall. Where jutting occurs, as with radial lays, this is the only place where wall gradation occurs and the stems, right at the end of the intrusion, curve in an upwards direction.

What kind of force can produce a variation of lays in one circle? If you are not confused yet, then perhaps this description of braided or plaited lays will help to make you so. This lay is just as described and has been seen in a few circles. Bundles of stems lie one over the other to form a typical plait or braid, the stems remaining straight in all cases. Some of the bundles have had two or more bundles laid at differing angles over and under them, so they

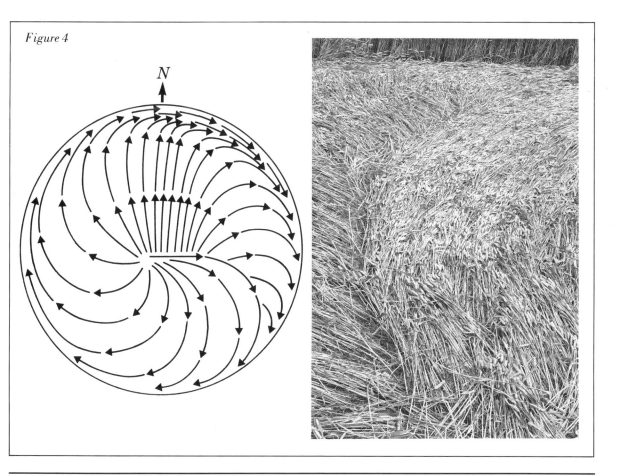

Figure 4

are actually intertwined. The force field that produced this would have to be operating like a knitting machine. On two occasions, in two different circles, some bundles of these braids have been laid in opposition to each other.

There seems to be no limit to the complication of lay that this extraordinary unknown force cannot produce.

A large number of circle floors are compressed hard to the ground, so that walking on them barely depresses them any further. By very carefully lifting some of the stems, it is possible to see their indentations left in the soil.

Circle centres have a large variety of patterns. Before 1986, they were usually the classic spiral formation, as can be seen in Plate 62, or a very close resemblance to this. The whorls at the centre start radiating from around quite a small patch of uncovered soil. They present a symmetrical hub from which they curve away to the circle wall and everything looks mechanically neat and tidy.

The variations on this classic centre encountered in the last two years have been amazing. There is the 2-metre-square central patch, where the stems are all facing in one direction.

Plate 62.

One side will have a straight line of stem root ends, as shown in Plate 63. From this line, and curving away at an angle, the spiral whorls commence. At the plant-head side of the square, the stems continue on around to form the whorls, the stems starting from the side of the square diffusing into them.

The 'parting' centre is interesting too. This is usually about 1·5 metres long and the whorls curve away from the straight central division. These 'spread' centres have many complications and are sometimes multi-layered too, in which case the true centre of the swirl is difficult to define.

The centre of a circle is usually the only place where curved and bent stems are to be found. One example can be seen in Plate 64. The stems of barley were wound around in a tight pot-like formation 1 metre in diameter. The heads of the plants were resting on their own roots. If the start of the central whorls is a tight curve, then the plant stems will be bent but not broken. Some buckling has been noticed in centres that have tight curvatures, but it has not been sufficient to destroy plant growth.

Plate 63.

Variations of centres range from a tightly wound pot-like formation to straight radials. All are compressed to the ground. Mixed centres occur sometimes: one half of the centre starts with a swirl, the other half is much straighter, even radial. Again, nothing can be taken for granted with centres, just as with lays. If it can be varied it will be.

We have used the word central because the start of the swirl is not always the centre of the circle. We have measured it up to 1·5 metres offset from the true centre. Sometimes the beginning of the swirl is hard to find because of what could be called a multi-start. The swirl appears to start in about three places in a square-metre area, and looks like a jumble of swirl beginnings, but all compressed, as usual, flat to the ground.

It would probably make solving the problem easier if it was obvious that the flattened swirl began at the centre. To make things more difficult, though, some central swirl whorls have been seen lying on top of whorls a little further out, so the outer whorls must have been depressed before the inner whorls.

Most circle centres suggest that the floor was depressed all at the same time because every bundle of stems has another bundle lying on top of it in such a way that the starting point is not clear. The bundles are quite distinct, as though gathered together by hand and slightly twisted, while being depressed and intertwined with adjacent and following bundles.

We have found the root end half of a bundle of stems covered by the head end half of a preceding bundle to form a Catherine-wheel

Plate 64.

Plate 65.

formation. No stems are bent or broken (see Plate 65).

An interesting centre can be seen in Plate 66, which has a number of the features in a collective manner, described above – bundles, whorls, swirls, parting and opposing lays. This photograph also includes a very good example of stems bent at the root end immediately after it comes out of the ground.

We have never discovered anything unusual in the soil at the centre, just an undisturbed surface. Sometimes there is some lichen or moss, if that is the general condition of the rest of the field.

Plate 66.

CIRCLE WALLS

The walls or outer edges of the circles vary from stark demarcation of horizontal and vertical stems with no gradation at all, to serrated and uneven edges that carry evidence of seemingly violent action. The edges of a circle that has whorls swirled two or more complete turns are usually neat and tidy. They look mechanically precise, with every stem laid perfectly and ready for inspection. All the way around, the vertical stems look as though they are standing guard over their colleagues, which are lying horizontally pressed down hard and flat to the ground. It is an astounding and unnaturally instant cut-off point.

Swirled whorls that have the shape of a French curve, an expanding arc spiral, usually penetrate slightly into the vertical edge. The end of the whorl will not be leaning against the standing crop, but will still be flattened, with maybe just a little gradation, like a roll of stair carpet that has been unrolled quickly, as though the end was flung into the edge. There will be several of these encroachments into the edge, depending on the size of the circle, and always between these penetrations the swirled crop will curve around the periphery. The penetrations are not equally spaced because the whorls are not equally spaced and likewise the depth of penetration varies around the edge.

In some circles, two or three whorls leaving the circle next to each other can cause a bay-like feature in an otherwise concentric configuration.

The greatest gradation of depressed stems takes place when the lay of the crop is formed radially. The swathes seem to blast their way outwards and even cause some damage. At a few sites where we have seen this peculiarity, stems are not only leaning into the wall penetration points but are broken. At these violent points there have been whole plants, roots as well, hanging upside down in the standing crop. Most of these extracted plants were not badly damaged, some not at all. There were perfect plants lying among the tops of the vertical growing crop.

Many of the blasted-edge circles had whorls commencing in a gentle curve at the centre, which gradually straightened out until they became radial at the edges, as in J in Figure 2 (page 123).

Not all multi-turn swirl circles finish up at the edge in a precise manner. Some contain a few stems leaning over at 45° or thereabouts. The lean is in the direction of the circle swirl and it seems as if the stems had only received half the induction, as it were, of the main flattened area. Stems affected remain in the same attitude until harvested.

Circular shapes have appeared in many forms. Only 2 per cent roughly of all the circles we have investigated are anywhere near a true circle. Most are ellipses, and not true ellipses either. Some are egg-shaped, or with a straight length on one side, or even two straight sections opposite each other. Where ellipses have 2 or 3 metres' difference in their axis, the central part will not always show this, for it is more than likely to have a concentric central swirl start. The swirl usually elongates in one direction to form an ellipse-like shape. The major axis of an ellipse is most likely to be in a north/south alignment. The significance of this is not apparent, unless it is associated with the earth's magnetic field.

It can be said that the smaller the diameter of the circle, the greater the chance of it being the shape of a true circle. At the same time, the smallest circles can still be depressed neatly and be very flat or have varying grades of rough lay.

The true shape of a circle can only be appreciated while looking down into it from an aeroplane. This is an unforgettable sight, especially when the sun is shining at a low angle.

The slope of the ground on which a circle is formed does not, apparently, influence the circle's shape. The creative force does not seem to slip downhill. There have been cases of ellipses where the major axis lies obliquely to the slope of the ground and cases where ellipses have been found on flat ground. Such is the contrary nature of this phenomenon that, having written these descriptions, next season may bring triangular or other shapes, as if to keep one jump ahead again.

Somewhere in every circle there is an unusual feature, which is perplexing when it is discovered. The basic phenomenon is puzzling enough without trying to reason why certain peculiarities are added as well. For instance, in a few circles we have noted small tufts and short narrow strips of standing crop somewhere in the circle. They could occur 2 metres in from the wall of a circle 18 metres in diameter or a short distance out from the centre. The standing plants are completely untouched and seem oblivious to what has happened around them. Whatever force induced the flattened plants to become depressed, missed the standing plants, ignoring them completely.

Some circles have contained a few single plant stems standing vertically in odd places around the floor. They are an obvious feature and very significant among the rest of the flattened crop, especially when on examination you discover they are emerging through a thick carpet of horizontal stems. Another unusual feature is a bundle of crop stems facing an entirely different way to the central rotational direction of the circle floor. These rogue bundles are always underneath the main swirl layer. Single vertical stems, broken off so that they are only 25 centimetres high, have also been found. About ten stems in this condition have been found in four circles and, as usual, for no obvious reason.

RINGS AND PATHS

Flattened rings add great beauty to a circle. As far as we know, they were first found in 1986; we have found no record of them before then. The width of the rings varies from site to site, and not in proportion to the diameter of the circle. A 17-metre-diameter circle with a 2-metre-wide ring has been found, as has a larger diameter circle with a 1·5-metre-wide ring. Such inconsistencies are characteristic of the phenomenon.

The band of standing crop between circle and ring also varies in width and again not in proportion to circle diameter or ring width. However, the direction of the flattened crop in the ring is always opposite to that of the circle, although that statement may well be incorrect after next season's display. The lay of the stems in the rings is similar to that of a classically laid circle floor, that is the stems lay tangentially to the centre. They tend to point slightly towards the outer wall of the ring. We have never discovered any contra-rotation in a ring. It is well behaved and lies neatly in one direction.

The inner and outer walls of rings show interesting characteristics, similar to the circle walls. The inner wall of a ring is usually well defined with no gradation. The outer wall will often be serrated slightly because of penetration of whorls, as in circles. These whorls are formed in the same manner as in the circle, but, of course, they are shorter and do not have much chance to divide. Some gradation does occur against the outer ring wall. It is most pronounced when the whorls are angled more sharply away from the circle centre. Again, there is the similarity to blast effect.

We have never found a ring containing a radial lay, they are always swirled clockwise or counter-clockwise. Some rings have contained 'bays' in the outer wall, and we have found no reason for this. The lay in the bays is usually flat to the ground and curves inwards into the bay then outwards again.

The shape of the outer ring normally follows the wall contour of the circle, but again there are variations, which are more prominent around elliptical formations. At one end of the major axis the ring is stretched or extended away from the ellipse wall. The band of standing crop is wider at this point, as is the ring width. In order to simulate the shape, an ellipse 'circle' with accompanying ring would have to be drawn on a sheet of thin rubber. One should then hold down the sheet across the minor axis then pull the sheet at one end of the major axis. The resulting shape is what is seen in the field, see Figure 3.

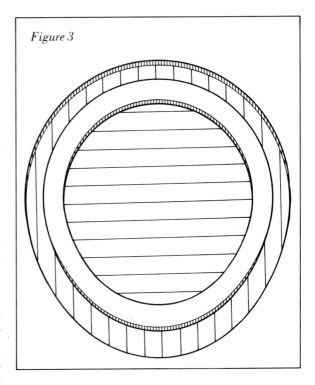

Figure 3

Walking slowly around a ring pathway, carefully noting unusual details, it is surprising how long the journey takes. After all, some of the larger rings have a circumference of 80 metres and there is often plenty to see in this length of ring floor.

Some rings also have a short pathway leading away from them. The floor is flattened in the same manner as in the ring. The stems at the junction of path and ring are a complicated affair of a circular swirl in the ring with an outflowing lay into the path. These paths do not go anywhere, but stop abruptly with some slight gradation, and are anything from 2 to 16 metres in length.

Occasionally circles have contained

flattened paths leading out of them, also going nowhere. One pathway leading out from a circle wall had a right-angled bend at its end which was about 1 metre long and ended abruptly. The crop stems at the end were resting against vertical untouched stems. The lay in the pathway was away from the circle, flat and unharmed, even the stems in the sharp corner were overlaid in a series of angles so that the path width was maintained. No stems were lying in the opposite direction.

From the air, the outline of a circle that has a path attached is similar to that of a tadpole, but without the tapered tail because the paths are always parallel sided.

As the lay in the path is always away from the circle, it seems to represent the exit rather than the entry of a force column. Whatever its cause, I feel it is very significant and probably holds an important clue.

Maybe one day we will discover a long, straight, flattened pathway in a corn field, unattached to any circle and remote from any tracks. After investigating all the phenomena so far, I am prepared for any shape or configuration to appear.

CHAPTER THREE

---◇---

CIRCLE DETAILS

Plate 67. *Upton Scudamore, 1987*

This classic central swirl shows typical whorls or veins, commencing at the centre and multiplying as they spiral out to the periphery. This example was in the central circle of a dice-like five-dot group. Close scrutiny shows the stems are not curved as they seem to be at first glance, but straight. The whorls look as though they were formed with a very large comb.

Plate 68. *Dog-leg Field, Longwood Estate, 1987*

Another example of a central swirl. The formation of the whorls and how they are swathed out can be seen clearly.

Plate 69. *The ringed circle, Punch Bowl, Cheesefoot Head, 1986*

This is an example of contra-rotation in one circle floor. The opposing lays can be clearly seen. On the left-hand stems the heads are coming towards you, while the stems on the right are facing away. The root end of each stem is bent over and pressed down hard with no damage to the plants, which is why they continued to grow and ripen horizontally. One buckled stem can be seen and one or two others are misplaced, which sometimes occurs when our feet catch a stem as we walk around the floor. The demarcation between the two rotations is quite remarkable: the stems have not been caught up and tangled or pushed sideways. Their transition from the vertical to the horizontal position in opposite directions was completed without gradation around the opposition line. Some slight splaying of the stems can be seen, which sometimes occurs where the end of a whorl or vein from the centre reaches the periphery.

Plate 67.

Plate 68.

Plate 69.

Plate 70.

Plate 71.

Plate 70. *Part of the smaller circle of the group of two, Bratton, July 1987*

This particular circle had another larger one superimposed on it a few days later. This photo shows a peculiar thin strip of contra-rotation, which several circles in 1987 exhibited. The direction of the swirl in this circle floor is counter-clockwise, but notice the clockwise flattened stems at the edge – E-type lay, with a Y pattern. When examining these particular circles, it looks as though the floor was first swept around counter-clockwise, then the edge was finished off with a thin circular stroke in the opposite direction. This plate also demonstrates the precise, sharp edge found in many circles.

Plate 71. *The Punch Bowl, Cheesefoot Head, 1986*

This bay was found in the north side of the ring round a circle at the western edge of the Punch Bowl. It is typical of those sometimes found in the walls of circles and the outer walls of rings. So far there is no obvious explanation for them. They are usually about 2 metres long and penetrate into the standing crop about 0·5 metres. Possibly they are the result of 'gap seeking' (see page 144), and occurred when the ring or circle was being formed. While forming the periphery, the creative force finds a bare or thinly sown patch and temporarily flows outwards, but the basic arc of the force is recovered, so a bay is left behind. Alternatively, bays may occur where the creative force began to increase the diameter of the circle or ring, but after a short distance the enlarging was aborted. Where two bays occur in a ring, the inside wall opposite never bulges outwards.

Plate 72. *Bratton, 1987*

Here flattened wheat stems, completely undamaged, are still growing quite healthily, which is typical of all circles and rings. The stems are bent sharply, close to the ground, and the rest of the plant remains straight. The radius of the bend is usually between 1 and 2 centimetres and if any buckling or damage is found, it has been trodden on at, or close to, the bend. To lift a patch of the stems upright again is extremely difficult. If you slide your hand underneath a bunch of plants and try to raise them, they feel stiff and reluctant. It is as though they have been induced to lie flat and that is how they intend to remain. If you try to reproduce what Plate 72 shows by whatever means you can think of, a high percentage of buckled and damaged stems will result. If the crop is laid flat by extra-terrestrial manipulation, then the force used is unique to us in its function.

Plate 72.

Plate 73. *Little Cheverall, Wiltshire, 1987*

The 'S' swirl here starts from the centre in a counter-clockwise direction. Half-way towards the edge, the swathes change direction and continue straight out, then, when they near the wall, they flick around to a clockwise direction. Many circles showed this peculiar lay for the first time in 1987.

Plate 73.

Plate 74.

Plate 74. *Punch Bowl, Cheesefoot Head, 1987*
This is a very unusual lay, in which the 'parting' central area can clearly be seen. It is also shown diagrammatically in Figure 4.

Plate 75.

Plate 75. *Ringed circle, Punch Bowl, 1987*
A section of radial lay with an underlying clockwise swirled outer band as described in the E type lay, X pattern. Obviously the outer band was laid down before the radial swathes: the force inducing the stems to flatten must have changed its direction dramatically. The end of the radial swathe at the top can be seen thrown against the wall. It was there that the underlying band terminated.

Plate 76. *Upton Scudamore, 1987*

This shows the result of either an indeterminate swirling force which began in a clockwise swirled direction, then changed to counter-clockwise, or a second visit superimposed an opposite swirl to the existing one. Parts of the circle floor showed good examples of both directions. As can also be seen, there are quiffs raised up where the clockwise lay was picked up and thrown in the other direction. Even with this dual direction, it was amazing how little the stems were damaged.

Plate 76.

Plate 77. *Headbourne Worthy, Winchester, 1986*

The stems in this tightly wound circle centre are curled so much that some buckling has occurred, about the only occasion it has been seen. As you will notice, the barley crop swirls away in a classic counter-clockwise rotation.

Plate 77.

Plate 78. *Punch Bowl, Cheesefoot Head, 1986*

This shows an average amount of central swirl for a densely sown crop of wheat. It is usual to find a bare patch at this point because the stems are induced away from it. It also shows another example of flattened stems bent sharply over at the root all growing healthily.

Plate 78.

Plate 79. *Bratton, Wiltshire, 1987*
The centre of a circle in a ripened crop of sparsely sown wheat. Because the crop was sparse the root ends of most of the stems can be seen clearly. Later in the year, after the field had been harvested, a circular green patch appeared. The grain had fallen from the horizontal, unharvested wheat heads and germinated. The hole in the soil was caused by a surveying pole used to measure the radius. This practice was rapidly dropped because it ruined the authenticity of the centre.

Plate 79.

Plate 80. *The Punch Bowl, 1987*
A wide indeterminate area which is more of a 'parting' than a swirl. It was found in the central area of the central circle in a group of five.

Plate 80.

Plate 81. *The Punch Bowl, 1986*
The wheat stems appear to wander away in a lazy manner in an area about 2 metres square. Outside this area they can be seen to pick up the classic spiral pattern. This was in the central area of the circle at the western edge of the Punch Bowl.

Plate 81.

Plate 82. *The Punch Bowl, 1986*
The neat and precise way most rings are created can be seen. Both the inside and outside edges or walls are well defined. The stems tend to point towards the outer wall, which is a feature of all rings.

Plate 82.

Plate 83. *Chilcomb, Winchester, 1987*
In this section of circle and ring is a classic example of 'gap seeking'. The ring swathe wants to follow the tractor tramline, recovers and then is made to conform to the natural curve; this has produced the wave effect.

Plate 83.

Plate 84. *Dog-leg Field, Longwood Estate, 1987*
Here some stems are flattened hard down while adjacent ones remain aloof and vertical. These features could be the result of inducing beams or columns not being concentrated enough to flatten all the circle floor.

Plate 84.

Plate 85.

Plate 86.

Plates 85 and 86. ***Thruxton, Hampshire, 1987***
These two photos show a well-defined circle.
The crop is not flattened and swirled, but a
jumble of buckled stems, most with the heads
missing. It is not the result of vortex because no
semblance of swirl was found. The stems are
buckled about half-way along their length. The
heads have been pulled down vertically with
minimum horizontal movement. The culprits
that created this circle are probably crows. A
farmer saw them in similarly damaged areas,
feeding on the barley seeds. Future investigation may explain whether the crows started the
damage and expanded it into a perfect circle, or
if it was created by some other means and the
crows exploited it once it was down and the
heads were readily accessible.

Plate 87. ***Winterbourne Stoke, Wiltshire, 1987***
This is a typical example of minimal gradation
at circle walls. The sudden transition of the
wheat stems from horizontal to vertical is one of
the amazing features of circle phenomena.

Plate 87.

CHAPTER FOUR

————◇————

PHOTOGRAPHIC AND MEASUREMENT TECHNIQUES

PHOTOGRAPHS

As with all new subjects, the photography and measurement of circles started in a basic way.

In the early 1980s, the few circles we visited were photographed from the nearest high ground or close to the edge of the circle wall. The results were quite good but not impressive because of the height of the crop and the height of the camera at eye level while standing on the ground. To look into the circle it was necessary to stand about 2 metres (maximum) outside the circle wall. Then, unless equipped with a wide-angle lens, it was possible to capture only a partial view.

It did not take long to realise that elevation was the answer. The various techniques we used to achieve this were hilarious at times. Holding the camera above our heads at arm's length became quite tiring, especially while waiting for a colleague to check that the lens was pointing down at the right angle; then, when everything seemed right, to discover we had not wound on the film.

To gain more height outside the circle, we made poles of various lengths with mounting brackets for cameras on top. The results were a great improvement, and much more individual than three or four of us standing in a circle with our arms stretched upwards. An outsider must have thought we were holding a ritual cere-mony. Personally, I favour the stepladder. I have a 2-metre one which fits inside the car and is portable. However, the reaction it causes at remote roadsides is comical.

At the Beckhampton site in 1987, Colin and I had parked on the opposite side of the road to a set of six circles in a wheat field. They were about 30 metres from the edge of the road. I erected my steps on a 1·5 metre bank opposite. From the top platform I had a fairly good, though somewhat oblique, view of the subject. As I peered through the viewfinder, I became aware of cars travelling at high speed along the long straight stretch of road, under and just in front of me. Most of them seemed to be driving on the far side of the road in either direction. Obviously they were giving a wide berth to this chap on a pair of steps, looking out across an apparently empty field. Not wishing to cause any incidents, I quickly dismantled the steps and ended the distraction.

At this particular spot there were obvious marks in the wheat field and damage to the wire fence near the circles, where a vehicle had left the road. I wondered if the driver had seen something that caused him to crash into the field, coincidental with the circles being formed.

There must have been some raised eyebrows on other occasions. After obtaining a farmer's permission to enter his field, two or three of us would march along one of the tramlines in Indian file, with me shouldering the stepladder, and Colin and Busty toting long camera poles through the waist-high crop.

We mostly use colour film because monochrome cannot reproduce the variations of colour in floor patterns. It is useful to record crop-colour changes as summer progresses and to show new green growth through the swathes. Colour also eliminates the need for consider-able explanations in many of the detail descrip-tions. The subject is displayed to the viewer as seen by the photographer. Monochrome is an asset for reproducing non-colour features, as in newspapers. We have used infra-red film for some aerial photography, but nothing startling has ever been disclosed in or around the circles. We had hoped there might be a residue of whatever visitation might have taken place, by machine or entity.

Choosing the best film speed is important for several reasons. As the subject is likely to be in a breezy location, general views are taken at fairly high speed to prevent blurring or swaying crops. Details found in the shaded part of the walls will require a fairly high ASA number. Low sun angle shots during early morning and late evening also should have a high number. A favourite is 400 ASA, which is a good general-purpose film. After taking about 200 photos using 100, 200 and 400 ASA, I have settled for the 400 ASA.

The Punch Bowl at Cheesefoot Head, near Winchester, has been the scene of many 'scrapes'. The amphitheatre is bounded on its west side by a curving road which makes its way uphill from Winchester. About half-way up the hill the road passes through a wooded area with a narrow copse on the left-hand side. It is through this copse that glimpses of the field, about two hundred feet below, can first be seen. Suddenly, the copse ends and the whole field comes clearly into view. Then the road begins a curve to the left with a raised grass verge between it and the boundary fence of the Bowl. Motorists suddenly see a number of people, with cameras and binoculars, all pointing down to the circles in the crop. Naturally, they cannot resist a peek over the left shoulder to see what has so much interest. Have you ever tried to hold a car in its correct line on a curving road without looking where you are going? That

is why the 00·5-metre high verge edge is slightly modified every time we find circles in the Punch Bowl.

From this vantage point the circles are very photogenic. Hundreds of photographs must be taken there by a great many people. We have met not only local people at this site, but many from other parts of the British Isles and from other countries too. Sadly for Mr Bruce, who farms this particular piece of land, a minority of the public walk down the side of the Bowl and haphazardly across the field to the circles, making destructive tracks.

The range of lenses used on the still cameras are from 150mm right down to macro to cover all requirements. Zoom lenses provide the most suitable choice of subject composition from hilltops down to ring-side views. The macro lenses are used for recording details of marks on stems, if the reason for them is not obvious.

The accessory bag holds many useful items including collapsible tripod, boxes of film, lenses, first aid kit, compass, maps, marker rods, 20-metre tape, clip board with writing pad, vacuum flask and sandwiches. When we set off to visit a new site, we are prepared for most eventualities.

Busty bought a video camera in 1985 and has shot many hours of film on the ground and from the air. The dramatic symmetry of the circles is seen at its best from an aircraft and good detail is picked out with the powerful zoom lens. Busty had two traumatic experiences with the video camera. The first was at the double-ringed circle at Bratton in 1987. He had an extra long metal pole, the end of which was fixed to the video camera, and had positioned himself about 30 metres down the slope, away from the outside ring. The rest of us were setting up equipment using poles and steps. Suddenly, there were shouted warnings mixed with a few curses. I looked around to see the video camera on its pole disappearing into the crop. Busty had managed to raise the pole by pushing it upwards from its centre, with one end planted firmly in the ground. He started panning around but the weight at the top decided to obey gravity, slowly at first, increasing as the angle of the pole changed. Despite his colourful instructions to go back up again, it finished up horizontally in the wheat. The video pictures give an exciting view, the sort a pilot would have dive bombing the circle.

A couple of hours later we set off for the circle at Town Farm. As we walked along a perimeter tramline close to the farm buildings, Colin was leading, I was in the middle complete with steps, and Busty brought up the rear. We had just rounded a bend at the corner of the field when a shout and a stream of curses sounded from behind me. All we could see was a patch of wheat thrashing about with unprintable exclamations coming out of it. Retracing our steps we came upon poor Busty, flat out on the ground, still clutching the video but minus the pole. He had tripped over a length of old fencing barbed wire, which Colin and I had stepped over. We helped him to his feet, dusted him off and recovered his equipment, but he complained of a very painful little toe. Despite his ordeal, he carried on, and took some excellent pictures of the circle. It had been a shock for us as well. We could almost be excused for daring to think he had been captured and drawn down out of sight by some force.

MEASUREMENTS

When three or four of us visited a site in our early investigative days, we organised the filming so that we each photographed different parts of the circle.

When we reach a new site, first we photograph general views outside the circle, then walk carefully across the floor to examine the centre and take pictures of it. We place a compass on the ground in the central area. One person holds the measuring tape about 0·5 metres above the compass, while another takes the other end and walks magnetically north to the wall. The measurement is noted on a prepared plan of a circle with the eight main compass points marked on it. The eight measurements are recorded, usually counter-clockwise because it is easier to hold the tape in your right hand against the circle wall if you are right handed.

Any salient features are measured from the same point and compass readings noted. The crop height is also measured, as are pathways, partings, blast effect bays and any standing stems on the floor. If the circle has a ring, the widths of the standing band and ring will be measured from the compass, covering all eight points.

If there is a group of circles, such as a dice-like five formation, the centre-to-centre distances and compass directions have to be measured. It is convenient to use magnetic north; true north can be transposed later, if necessary. Included with the radial measurements are the number of turns of the swirled swathes and their pitch distance, which sometimes increases as the swirl moves outwards.

Measuring centre-to-centre distances of circles often means walking back to the edge of the field and then out again along other tramlines to prevent damaging the crops. We throw the tape's case from one circle to another once we have pulled the full length of tape out. We also measure distances to pylons, power-line poles, paths and tracks if they are within a reasonable distance, which may bring to light some significance during collation.

When circles in a group are spaced well apart, leap-frogging with the tape – using a scratch mark in the soil or a stick for each tape length – is necessary. Long nails are very useful. At first we used sticks, but found they were useless in rock-hard soil. Nails can be hammered in with a stone, but should have a short length of coloured tape attached for obvious reasons. I learnt the hard way, and had the saying about the needle in a haystack brought home to me with a vengeance.

The central area often contains complications, such as two initial swirls, a parting, a herring-bone pattern, or a swirl and a parting. They are all measured with care so a drawing can be made later, and the compass is used so the central features can be aligned correctly with the other information.

Notes, descriptions and explanations accompany the measurements because often only one visit to the site is possible. Not wanting to strain relations with the farmer, we gather maximum information in minimum time.

We each have our own allotted tasks to ensure there is not much duplication of information. Notes, measurements and so on are swapped afterwards in the local pub. I often wonder what inadvertent eavesdroppers must think when they hear us talking of 'swirled swathes', 'blast effect', 'contra-rotation' or 'twenty metres centre to centre', and see drawings of peculiar symbols, circles and signs. They could not be blamed for thinking a new religion had come upon them.

CHAPTER FIVE

THEORIES

How are these circles and rings created? Whenever the subject is discussed, this question arises. It is posed in numerous ways, especially by newcomers to the subject: 'What do you think it is?' 'What makes them?' 'How do they get there?' 'Is it a hoax?' Then from the serious thinker: 'Do you think it is a magnetic force?' 'Is it electro-magnetic?' 'Is it an anti-gravity force?' 'Could this be the result of a piezoelectric effect . . . micro waves . . . ultrasonics . . . an atmospheric sourced vortex . . . a terrestrial sourced vortex . . . a confluence of the last two forces?' After eight years of investigation I am expected to know or at least have a pretty good idea of what is involved. When I reply, 'I just don't know how the circles are created,' an air of dissatisfaction permeates the atmosphere, making me feel guilty. Then the stereotyped suggestions are put forward, which is probably satisfactory, because at least they trigger debate. It is to be hoped that when this happens, people will become sufficiently interested to recall other circles they may have seen or heard about.

It is perfectly natural to ask if the circles are hoaxes, but very difficult to explain why they cannot be hoaxed satisfactorily. The first and favourite method suggested is to drive a stake into the ground, attach one end of a chain or pole to it and pull the other end around in a circle. This does not work. The crops on the circle floors we have examined are pressed hard to the ground and in 95 per cent of cases are completely undamaged; no breaking, no bruising or buckling, no heads broken off and no holes in the centre.

To find out what has to be done to make a flattened circle by the above method, I carried out the following experiment. Using the edge of a standard wheat field, I placed a wooden rod, 1 metre long, on the ground between the stems. To the centre of the rod I attached a cord, 1 metre long, and to the other end I tied the hook of a 10 kilogram maximum spring balance. I pulled the spring balance horizontally until the maximum reading was attained. The rod had moved forward about 30 centimetres and upwards at an angle of about 45° I could not keep the rod moving horizontally. The stems initially pulled over by the rod leant against adjacent stems which moved over and leant against more stems and so on. This gradation forced the rod upwards and to press it down to within 5 centimetres of the ground required a pressure of about 15 kilograms. When these figures are extrapolated to produce a 10-metre radius, a pole with a weight of something like 150 kilograms will be needed to keep it on the ground. Enormous pressure would be required to pull this mass around in a circle. Even if this was achieved, the stems would be badly damaged, especially by the feet of the puller, who would have to be in front of the pole. Whatever variations there are to this theme, the details described earlier could never be produced. Imagine the weight of equipment required to carry out this operation.

Some circles have been found away from the tramlines made by farm tractors. Readers may be wondering why the tramlines are there. During the crop's growing time, a tractor fitted with a 10-metre-long spray boom on either side traverses the field so the whole crop has a dose of insecticide or fertiliser. The tractors are always driven down the same tracks to keep the destruction of the crop to a minimum. The circles with diameters smaller than the distance between sets of tramlines have no tracks leading to them. So heavy flattening equipment would have to be airlifted in. I have tried to walk into a field of wheat without leaving tracks and it is impossible – picking your way through the stems without breaking any can't be done.

We have visited two hoaxed circles, and it was patently obvious that they were not the natural type. One, which was made for publicity and possible financial gain in a demonstration by the media to show how it was done, was a hopeless mess because of the method used. One person stood in the centre holding one end of a rope, while two others holding on to the other end trampled the crop down with the sides of their boots. Having trodden down the perimeter, they circled inwards to complete the flattening. The result was a roughly swirled and broken-stemmed floor. In the other hoax two youths lay on the ground and rolled over and over, directed by two other youths standing in the centre. They left a roughly diamond-shaped area with no specific central swirl. The floor had some flattened areas and bundles that looked like sheaves that had fallen over.

Another hoaxed circle was recently shown on television. A line of men were holding a rope to keep them in position. The man on one end remained stationary while the others walked radially around him. This formed a flattened circle, but the crop was badly damaged and none of the subtleties described elsewhere in this book were visible.

The real circles are objects of beauty, precision and mysterious detail as described already.

It will be realised now how difficult it is to simulate a circle created by a force field, if it can be described as such. Many experiments have been carried out to try to flatten crops in the way seen in these circles, but it is impossible to do so without inflicting damage on the stems and leaving marks in the soil. To create spiral whorls, swathes, lays, multi-layering, contra-rotation, precise sharp edges, in most cases without entry or exit paths to the circle, must require a special kind of force.

After extensive enquiry and prolonged personal observation, we have no evidence that these circles are created except at night. We know they can be created between early May and late September in southern England. Within this period, they are independent of the date, day and weather conditions – heat or cold, dead calm or windy, dry or rainy. They have no apparent connection with the position of the moon or planets. Geological location – flat or sloping ground, the presence or absence of trees – seems to have no bearing on where they are formed. They have been created in bowl-shaped localities or on the tops of hills under all these conditions.

In 1986 Don Tuersly and I decided to carry out all-night vigils at the Punch Bowl, Winchester, a site where circles have appeared for many years. We started in the second week of June and continued until 6–7 July. That night I stayed until just after midnight when Don arrived to relieve me. With the aid of good light-gathering binoculars, I knew that nothing had occurred up to about 11 pm and then it was too dark for observation. I arrived back at the site at about 3 am. The night had deteriorated from being dry and cloudy before midnight to a fine drizzly rain. At 3.45 am, the light rain had eased to just a few spots, and by the first dawn light I could see something had arrived in the field. The dark shape of an ellipse was way out in the centre, about 300 metres away. From where we were above the Punch Bowl, we had to wait a further fifteen minutes to be able to confirm that a circle with a ring around it had been created. We had neither heard nor seen anything at all while very carefully maintaining our vigil.

Many other confirmations of night-time creations come from farmers and people living near circle sites. 'It wasn't there last night, but I noticed it first thing this morning,' has become almost a stock statement. The evidence is overwhelming that circle creations only occur at night.

What force registers its presence in crops under any of the foregoing conditions – a force that leaves no trace of any kind other than the circle itself?

From the evidence we have, it would seem that the flattening, swirling, whorling and swathing takes place over the whole of the circular area at the same time. This unknown force creates all the various circle floor details in less than half a minute but probably more than five seconds. We conclude this because to achieve a swirled, flattened condition pressed hard to the ground, the stems must have a maximum vertical to horizontal transformation time, above which they would be damaged by whiplash. Once all the stems had started moving in a steady manner, the whole circle, whatever its size, would be formed in possibly twenty seconds.

There are many possibilities concerning the way in which this force presents itself. The formation of some circle floors suggests rotating bands of force, about 0·33 metres wide. They would be numerous for the large-diameter circles and correspondingly fewer for the smaller-diameter circles. The bands would be as though they were hanging vertically in strips to form a circular column. The column may rotate at different rates to form either the multi-turn whorl spirals or the lesser-turn spirals. Of course the bands may remain stationary and thus form the straight radial swathes that have been described. They may emanate from a pointed source and spread out to form a rotating conic spiral. The apex of this cone could be 10 or maybe 100 metres above the earth. It is estimated that this variation of cone height would be able to create circles of varying diameters.

Unfortunately, things are not that simple and to keep guessing at various ideas of how crop manipulation takes place is fraught with obstacles. For instance, just take a small diameter circle that has been flattened and swirled in a neat and precise manner. It may not be too difficult to imagine various ways a force field could do this but, as we know, the problem is not so straightforward with all the variations now documented. One significant obstacle among the many challenging our intelligence is a peculiarity I have termed 'gap seeking'. When the force that creates these circles is operating in a whirling rotary motion, it is diverted from its arc-describing path outwards into any plantless gap that happens to be adjacent to the force band. The irregularities occurring in the walls of circles and rings are mostly caused by this gap-seeking element. This phenomenon is illustrated in Figure 5.

From above the form appears to be a circle with a flat side. A deviation will occur where the crop has a sparsely sown patch and the amount of deviation depends on how sparse the patch is. Some poorly sown crops display a variety of densities. When a circle is created in such a field, its wall will take on a serrated shape, due to this gap-seeking quality. The impression given is of a force that can be temporarily distracted from its circular path but under sufficient control to return to the true intended line of arc.

Of course, this gap-seeking characteristic must occur all over the floor area of the circle during its creation period and it is probably because of this we see various oddities in swirls and swathes, where the lay of stems are deviated away from the general swirl direction and then return again.

An evenly sown dense crop that has received a flattened swirled circle shows a much more disciplined control in the operating force. The wall is precise with practically no half-way flattened stems or serrations, and where a

Fig. 5   The result of 'gap-seeking' force when its circular path is 'interfered' with by tractor tramlines. The dashed lines show the true line of circle and ring walls. After the walls have

deviated, they are pulled back on to their correct line of arc. The whole distance of a deviation and recovery varies from 3 to 4 metres when these occur in a circle 20 metres in diameter.

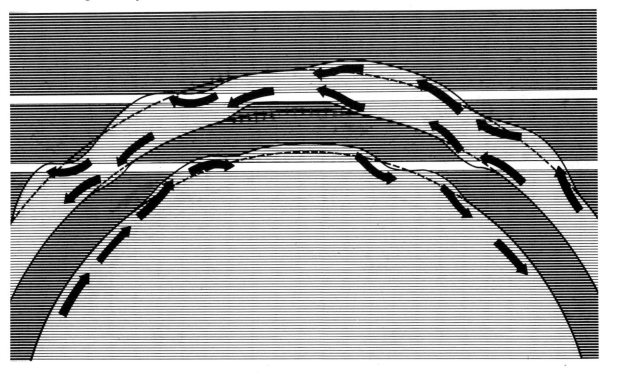

tramline gap is encountered the deviation takes place over a shorter length. There is a short, sharp deviation and recovery.

The walls of circles formed by blast-effect radial swathes vary with the density of the crop. In a dense crop a short upward curve is made, but a more leisurely curve appears when the crop is sparse. This is where the horizontal gap-seeking effect in a swirl changes to vertical with the radial swathe.

Any specification for a force that can produce all these complications will have to include the following features: flattening crop stems of various thicknesses hard to the ground without damage, spiral whorl rotation, contra-rotation, multi-layering containing multi-direction lays, blast-effect radial swathes, and all the different central area formations. Not only must the force be able to lay the crop in either direction of rotation, it must be able to do both rotations without gradation on the floor of the same circle.

It must be a strong force of short duration that induces horizontal growing into the plant, replacing its natural tendency to the vertical. Once these plants have been flattened by this force the head end never attempts to grow vertically again. The only exception occurs when the stems are only partly flattened to the ground. It appears the force can vary in strength and only cause a 40° to 60° angle of depression, and then, aided by the wind and some tendency to grow upright, the stems may become almost vertical again. When this unusual floor pattern is examined, at least 25 per cent of stems in the circle will be well flattened. This raises the question of how the force can be so selective. It must also be able to make the root end of thick-stemmed plants pliable enough to bend to a sharp, almost 90° angle, without fracturing and disturbing the plants' growth rate.

It must be able to construct a flattened ring around the outside of a circle, closely following the contour of the circle wall. Besides creating circular shapes, it must also be able to flatten a dead straight pathway several metres long. It should have the ability to miss narrow, arc-shaped areas of crop and so leave these stems standing, like a low, slim, curved screen. It must be so violent that some plants are pulled up randomly or ejected from the soil and thrown into the peripheral standing crop. It must be quiet. With the exception of two possible cases, which will be dealt with later, there have never been reports of sounds associated with the appearance of circles. It must have no lights or illumination associated with it.

In a nutshell, the force's specification requirement to record a flattened, swirled circle in a receptive crop is this: a silent, short-duration, strong, contra-rotative, damage-free flattening, swirling, whorl- and vein-forming, swathing, stem-bending, horizontal-growth-inducing, non-growth interfering, straight-path forming, plant-extracting, total darkness operating, gap-seeking, superimposing, circle-group forming, weather condition free, extraneous marks free, topographically conditionless, worldwide operative force. The inclusion of most of these twenty-one requirements is necessary for all circles.

It is very sobering to stand in one of these circles and ponder what force could have arrived and departed, leaving behind this beautiful record of its visit with no clue as to how it was achieved.

The descriptions and details given so far may have given the reader the impression that we think the creating force of the circle operates only from above the earth's surface. Everything described could, of course, be the outcome of a sub-surface force, with a pulling down instead of pushing down force. It has to be one or the other, or even a confluence of both. This last would probably require a special condition whereby the forces from below and above coincided precisely.

A horizontally acting force is thought to be least likely based on the evidence of how plant stems, in some circles, are depressed into the soil surface. The impressions of the stems can readily be seen when the bottom layer is carefully lifted. A truly horizontally acting force might induce the stems to lie flat, but they would not tend to go sub-surface.

Another reason for considering atmospheric or subterranean force columns is the ellipse-shaped 'circles'. A round column, when projected through a flat plane at an angle, will describe an ellipse. Obviously this occurs, as we have seen, when a vertical column meets sloping ground.

Some 'circles' have been oblate, like a stretched circle, where opposite sides contain a straight length of wall. To the engineering mind, this shape is obtained by a vertical, round column being moved horizontally over a flat plane. To produce the swirl, the column would have to rotate and contra-rotate. To create a contra-rotated flattened circle, a clockwise rotating force column is necessary. Fitting tightly around this column is a counter-clockwise rotating sleeve or tube of force. These rotating forces are contained in such a manner that they often display no gradation whatsoever. There is no dissipation of the force at the column's peripheral surface. All known forces have a field which, in some instances, can only be contained and prevented from giving extraneous effects by some form of metal shielding.

The precise vertical walls of some circles could lead the observer into imagining that a tubular shield of some form is lowered into the crop, inside which the swirled flattening takes place.

The next step is to consider this same force creating a circle that displays the blast effect swathes from, as it were, the hub to the rim. This is not a rotating force; it commences at the centre and radiates outwards. An analogy may be drawn from imagining a jug of water being poured on to the centre of a flat, round dish, the water spreading rapidly outwards until it hits the sides. It is no longer a rotating force, but one that is fed in at the centre and maintained until the length of the circle radius is determined, at which point it terminates with one upward sweep.

The force that produces the swirl-ended radial must be similar, only in this case something happens near the end of the radial blast effect which causes the straight swathe to swing suddenly to one side, as though it had come up against a shield.

It has already been difficult enough to guess intelligently how this force functions. Now consider the S-shaped swirl, which contains parts of all the shapes so far discussed. It starts off as a swirl at the centre, straightens out to form a blast radial, then swirls around at the end, against the wall. It has induced swirl, blast and reverse swirl over the whole circle-floor area. There is no way any known force can be controlled to produce the complications of this phenomenon.

Many people have suggested that the rings we describe are caused by a virus. Only when they are shown photographs of circle floors, or visit a circle, do they realise that under no circumstances could a virus produce such a phenomenon. When a virus affects a crop, the crop's growth is impaired or distorted so the stems buckle or lean over in a random manner.

The well-known fairy rings, which appear on lawns and regularly mown grassy areas, have nothing to do with cereal crops. They are really parasitic fungi known as *Marasmius oreades* or *Clitocybe gigantea*. They also appear in grazing meadows and we have taken photographs of them in order to dispel any confusion over these and our subject (see Plates 88 and 89).

I have checked sixty books dealing with fairy ring fungi and other associated fungi and viruses. Not one mentions the formation of flattened swirled circles in grass or any other crop. A fairy ring is a ring of lusher, darker

Plate 88.

growth than the surrounding crop, whether the crop is inside or outside the ring. The ring often remains for a great many years and expands continuously.

Electro-magnetism is frequently suggested as the motive force to create these circles. How it can be applied to plants to cause a neat, flattened circular swirl as well as all the other forms is hard to imagine. Electro-magnetism is produced by a coil of wire, usually copper, which has an electrical current passing through it. The electro-magnetic field thus formed will attract a mass of ferrous metal to its core. The only way this force can be applied to plant stems is for particles in the stems to become charged in such a way that they are induced, with considerable force, to flatten to the ground. The stems must not go down in a random way but in the orderly swathes described. Experiments with electro-magnetism on plants have shown that under certain conditions blades of grass can be made to move but not to flatten down hard. Imagine the amount of equipment that would have to be carried into a remote field in order to create a 20-metre diameter circle.

Plate 89.

Simply, electro-magnetism works as follows. A straight conductor carrying an electrical current will have a spiral magnetic force field around and along its length. A coil of wire carrying an electrical current will have a magnetic force field through its centre, returning around the outside of the coil to complete the circuit. To construct a flattened, swirled circle by some form of electro-magnetic force a conductor would have to be placed vertically on the ground or protruding from the ground. If this conductor, while carrying a DC current, was capable of having its diameter increased, then maybe the crop stems could be induced to lie flat. To create a circle with a ring around the outside, the circle could be swirled clockwise with the current flowing along the expanding conductor in one direction. The current would then be switched off to leave the standing band. With the conductor's diameter continuing to expand, the current is switched on, but flowing in the opposite direction to create the opposite rotation for the flattened ring's swirl. It can be seen this theory is capable of creating many of the configurations so far discovered.

For blast-effect floor patterns, a device resembling a large coil of wire with its core in a vertical position would be needed. The magnetic force field's lines exiting from the end of the coil would push the crop stems down and away from the centre of the core.

To carry this theory a step further, consider the S-swirl floor pattern. If the coil in the previous paragraph were to be rotated counter-clockwise momentarily and then clockwise with the current flowing, it is possible that the crop would be laid flat with an 'S' swirl.

With these basic principles of electro-magnetism in mind, the reader may now be able to expand them and relate them to various sources that may come to mind. The source would have to simulate a vertical conductor and a coil with a vertical core.

It must be emphasised that one circle cannot be used as a typical example for applying theories. Whatever the force is thought to be must be capable of carrying out everything that has been described as found in the circles.

Biological interference, rutting deer, helicopters, pole and chain hoaxing devices and anything else you can think of have all been suggested. Not one suggestion gets very far when it has to produce the effects we have seen.

Magnetism is a force that is not fully understood. We all recognise it by its unique effect on ferrous metal. Some people think it plays a significant part in forming a circular flattened crop. Just how is difficult to imagine. A magnetised plant stem, if that were possible, would be polarised either north or south at its head and the opposite at the root. If all the heads were polarised either north or south, they would all remain parallel to each other while in a vertical magnetic field. It would be impossible for this field to lay the stems horizontally in a circular pattern inside a vertical non-magnetised wall of stems.

The theory that all these circles appear over archaeological sites also seems very unlikely. Plenty of literature is available describing how crops are able to display circular shapes, which are only visible from above, as on adjacent hillsides or from an aeroplane. The shapes are visible because of two conditions: against the general colour of the crop, they reveal themselves as rings of either lighter or darker hue.

The reason for this concerns ancient, circular building foundations. The buildings were abandoned and subsequently became derelict. The stones from which they were constructed were either used for other constructions or scattered with time. Depending on how many stones were removed with respect to ground level, a shallow- or deep-ringed trench would be left. As time passed, soil would fill the trenches through the action of wind and rain. The shallow trenches consequently provide a shallow depth of soil with probably a stone underlay, thus reducing the supportive bulk of soil that has excessive drainage. The result is a ring of weak crop which is shorter and paler than the main crop. Deep trenches will have filled with blown and washed soil, forming a ring of richer earth. This will provide a rich ring where the crop will grow taller and darker than the main crop. There are slight variations to these explanations, but the principle remains the same.

The precision with which some single circles are made is only a part of the problem we are trying to unravel. The groups of circles add a further dimension to the problem. Consider a group of just two. Typical diameters for these would be 8 metres and 4·5 metres and the space between their rims about 2 or 3 metres. The floors may be swirled in the same direction or contain opposite rotation. Did two force columns create these simultaneously, or are they the result of one force field leaving two impressions that have differing parameters? If there were two independent force-field columns, they must have operated completely independently of each other and with disregard to rotational interference. Dual circles found with precise standing edges show no sign of repulsion or attraction at adjacent edges. The foregoing observations are valid for groups of three circles in line and five circles in dice-spot formation. Trying to consider where the source of the force field emanates from is just as enigmatic as everything else concerning this subject.

Just supposing the force is earth sourced, the questions arising from this would be: what force? How deep? Is there a network? What determines the rotational direction? And what determines the exit points?

What are the forces known to exist inside the earth? Well, of course there is 1) gravity, a good starting point considering we are dealing with flattened crops; 2) piezoelectric generation; 3) gaseous pressure build-up; 4) electric potential differences between various elements; 5) mechanical pressure caused by sliding rock strata; 6) centrifugal force caused by the earth's rotation; 7) tidal effect caused by sun and planet coincidence; 8) volcanic pressures; 9) steam pressure.

Probably more forces could be mentioned, but it is highly likely they would be related to one of the above. Briefly touching on each one may spark off a train of thought the reader might like to dwell on and maybe research into.

1 *Gravity*. Good old, taken-for-granted gravity. Certain scientific phenomena excepted, everything is attracted downwards perpendicular to the earth's surface. Cereal crop stems, for example, remain upright because their roots hold them in that position. If the stems were suddenly parted from their roots, they would fall in a random manner. To persuade these stems, attached to their roots, to fall in an

orderly rotated swirl, gravity would have to be manipulated to contain a horizontal element. This would be necessary to cause all the plant stems initially to lean in the same rotated direction about a central point. Once they were all leaning over the same way, then perpendicular gravity could pull them down flat. Of course the force of gravity carrying out this act would have to be increased many times to compress the floor layers to the extent they are found.

2 *Piezoelectric generation* is a well-known method of deriving a variable electric potential from certain crystal elements when they are subjected to pressure. It follows that this might occur in crystalline rock structure being subjected to pressure deep inside the earth. Perhaps there is some way in which the ensuing electric potential output can build up a charge, which is stored and discharged out through the earth's crust, recording its exit by interfering with standing crops. How this interference is induced is a major problem.

3 *Gaseous pressure build-up*. There could be a peripheral relationship with piezoelectrical generation here. High-pressure gas pockets may cause rock strata to become unstable, which, in turn, creates a knock-on situation culminating in a circle-creating force.

4 *Electric potential differences* occur between certain dissimilar natural metals located in the earth's structure. An electrolytic substance between these two metals affords a circuit path for voltage to flow. How it would be accumulated and discharged is a matter for conjecture.

5 *Mechanical pressure* caused by rock strata movement could have a relationship with piezoelectrical generation.

6 The bulge caused by *centrifugal force* at the earth's equator could be related to gravity and piezoelectrical generation.

7 *The tidal breathing effect*, caused by the gravitational influence of the sun and moon on the earth's crust may also be related to piezoelectrical generation.

8 *Volcanic pressure* can probably be disregarded as a direct cause, even though there is a possibility of far-reaching piezoelectric effects being distributed through the earth's crust. If this is so, accumulated charges may find discharge points under certain conditions. These discharge points would probably produce clouds of dust in a minor explosion. They would be unlikely to reproduce the fine details of the circles.

9 *Steam pressure* from within the earth may be bracketed with volcanic pressure in producing similar results.

Maybe somewhere in this list of the earth's forces is the clue we are looking for. Whatever it is, and however it is derived, it seems to be intelligently controlled (see Figures 1 and 2, pages 118–119 and 123).

Most readers will be familiar with the term 'ley lines'. Some may have practised the art of dowsing with varying degrees of success. It seems without doubt that detectable lines of some unnamed force exist and are traceable along either long or short distances by various forms of dowsing. This force, which is probably in the form of a network, may have static and variable lines. The variable lines may move in a wavelike motion, like a long pennant in a breeze or a frond of water weed in a gently flowing stream. Because of this movement, there may be random or ordained intersection of lines. Should this occur under specific conditions, a short-lived rotating spiral force may be formed with sufficient energy to create a flattened circle.

Ley lines may be considered as a surface force, and there are others that can be put in this category.

Birds, mainly crows and pigeons but especially crows, are known to create a circular formation in a cereal crop. Plates 78 and 79 are a good example of a crow-damaged circle. The nucleus of these formations may be a small patch of wind damage, as described by one farmer, or animal damage, such as a deer lying down, according to another farmer. The crows will drop down into this depression and eat the young seeds lying on the ground. They are far too heavy for standing stems to support them while they strip the seed heads. Once they have a favoured depression, they can now flap up and grab a seed head of a standing plant and bring it down, usually buckling the stem half-way down. They continue, in their dozens, pulling down the stems and increasing the circle diameter to as much as 17 metres. Of course, shapes other than circles are formed. Nevertheless, an almost true circle has been photographed that was created by crows in this manner.

Animals, especially deer, are sometimes given the honour of creating these precise circles. Many times we have seen their erratic tracks across cereal crop fields and on several occasions have actually seen the culprits moving away in leaps and bounds. When we approached the ringed circle at Chilcomb for the first time, Colin and I stopped 50 metres away to take photographs. I noticed a dark brown head with large ears looking at us over the edge of the ring. I tried to photograph it, but the deer bounded away down one of the tramlines. A minute or so later a second deer leapt from the circle into the ring and raced away to join its partner, after which both of them bounded in a zigzag manner to the far side of the field and disappeared. There was no obvious evidence they had occupied the flattened area for very long and careful inspection brought to light no droppings or recent damage. There was no way they could have been responsible for the circle's appearance. Just because they have been seen in circles, some people must assume they have created them instead of using them as rest areas.

One example of how reluctant people can be to accept that mysteries occur arose when I called upon a retired county council byways maintenance man in 1983. I was told that he had seen the circles for many years while carrying out his job in the local countryside. During our discussion he told me, 'It must be fourteen or fifteen years since I first noticed them. They are made by rutting deer, because I have seen deer in the same field as the circle. I think the deer run around and flatten the wheat.' Although I suggested that this would damage the crop, he was unshakeable that anything else could cause them.

We have also seen rabbits in the circles, but I for one am quite convinced that they were only doing what rabbits do in any field.

When a phenomenon like the circles occurs, it is natural to wonder who or what did it and how. Basically, humans do not like mysteries, they make us feel uncomfortable and we like to have a neat scientific answer for every problem. Some people need to make everything conform to our conventional sciences. However, I find it arrogant to think we know all there is to know and there is nothing left to add to our scientific knowledge.

Our purpose in this book is to present the facts, facts that you can verify by walking into a circle and checking for yourself. To say we have no opinion on what is happening is not entirely correct – in our opinion something is happening that we do not understand.

Next we must consider atmospheric forces such as lightning; winds; solar energies; the earth's magnetic field; thermal changes; static electricity, and other unknown forces.

1 *Lightning* is usually quite destructive. An explosive bolt of very short duration, capable of splitting trees and buildings, is not uncommon and the result leaves no doubt about the ferocity of its concentrated power. This type of power is hardly conducive to laying crops' stems firmly but gently to the ground.

2 *Air*, unless mechanically contained, is unruly and requires a supervision to perform in a precise manner, especially on a cereal crop (see Plate 90).

Much has been written trying to explain how it is possible for a whirlwind to create circles.

Apparently, a set of conditions is necessary to perform this unique task. First, there must be an escarpment, whereby the hilltop is 100 metres or so above the field containing the unsuspecting crop. The wind must be from a direction to blow over the hilltop, swirling, tumbling and dropping down in such a way as to create a vortex, because of the change in pressure and volume over the field area where it can descend on the crop. The whirlwind must now remain stationary while it gently flattens the crop with the swirls and patterns described and not suck any of it up, as whirlwinds do normally. They are not responsible for the few plants that are found thrown out and up-ended

Plate 90.

from the crop; whirlwinds are not selective. Unfortunately for this theory, many of our circles have been in locations remote from any hills.

We know that a whirlwind is caused by air moving in from every direction and meeting tangentially at a central point. The core spins upwards and is replaced by the incoming outside flow, mostly from the base. This is true of either ground- or cloud-initiated spinning columns. The cloud-initiated columns are mostly associated with water spouts and cyclones and are well known for their destructive power. Hot- or fair-weather whirlwinds are well known in hot countries and occasionally occur in temperate localities as well during the daytime, but never at night.

Anyone who has been near any type of whirlwind will know how noisy they are: even small columns in hot countries have a considerable roar. I have seen more than twenty columns at one time in the Australian desert and they always drift along with the prevailing air mass. Each one is carrying aloft sand, dirt, dried plants and other debris, proving their action is to suck whatever it can while spinning upwards. I have even seen an empty forty-gallon oil drum lifted up. The action of a whirlwind is analogous to a bath emptying. Imagine the surface of the water is ground level. Now invert it so the spinning column is going upwards and you have the whirlwind action.

It can be seen from the foregoing that whirlwinds cannot be responsible for creating our circles and rings, especially the latter. It is an insult to anyone's intelligence, after considering all the facts, to be asked to believe these beautiful, precise circles are created by a mass of swirling wind. It has even been suggested that five precisely positioned whirlwinds have created the marvellous five-dot dice-pattern groups. That suggestion is a joke really, another case of trying to make a theory fit the facts.

3 *Solar energies* include heat, spectrum-colour energies, charged particles in the solar winds, X-rays and all the other rays, photons and other energies yet to be discovered. As circles are, apparently, always created at night, the sun cannot be considered responsible, based on our evidence at the present time, unless some part of its energy is delayed to operate at night.

4 *The earth's magnetic field* may have an undiscovered influence on crop stems. It would be a terrific breakthrough for modern technology to understand how to manipulate the earth's magnetic field to create precise, flattened, contra-rotated, swirled circles. It may require an aberration in the earth's magnetic-field flow lines to cause a conic spiral which, due to its unnatural formation, is capable of plant manipulation.

5 *Thermal changes* must originate with the sun and possibly from the earth's core. How these changes can produce the kind of force necessary to produce the circle phenomenon cannot be imagined, but the whole subject is a mystery anyway so maybe the force does originate from what we think is the most unlikely source.

6 *Static electricity* is a firm favourite with many people and why not? However, it is difficult to imagine how many of the parameters

in circle construction would be met. Positively and negatively charged areas of plants would have to be carefully planned before the action. A horizontal force is needed initially to attract the stems in the direction of the intended swirl. What would be happening to the charges when the stems begin to slide down neighbouring stems is not easy to imagine. It is especially difficult to comprehend the charge patterns where contra-flow layers occur. To attract the plants downwards, a neat circle of very strong opposite charge is necessary in the soil. So both horizontal and discreet vertically charged areas have to operate in the correct sequence, otherwise random collapse would take place.

The main question has not been considered yet. What is the device that induces a force to act on the plants in the way we have studied? Many circles are formed during rainy nights and wet stems may not be compatible with static charges. The stems around the edge would have to be treated with anti-static fluid. This would be quite a task and much trampling would take place, which would play havoc with the crop stems. Of course, this is light-hearted conjecture and is included to highlight the fact that whatever conventional science is suggested to create these circles cannot be carried to a conclusion. There are too many parameters that must be included and cannot be met.

7 To apply any conventional science requires a considerable amount of apparatus. The pristine nature of the crops surrounding the circles when first discovered suggests there had been no ground contact by any human or any mechanism. Imagine setting up apparatus in the middle of a field of wheat on a dark, rainy night, without lights and making no noise. You would first have to break into the edge of the field to find a convenient tractor tramline. To walk along one of these in the dark, carrying your equipment and not stumbling into the edge of the tramline, would be impossible. This is only entering the field. How much more impossible it would be to do anything more complicated like producing a pattern like H or L in Figure 1, pages 118–119.

We have to look for a force that takes place independent of ground contact. This means we may be looking for an unrecognised force that is already in existence in some places but requires control and manipulation; or perhaps a transportable force used at will by some controlling power. Both ideas suggest an intelligence wanting to produce these manifestations for some unknown reason.

Members of the public who are fascinated by these circles appearing in the more readily visible sites and who speak to us there, often come out with a straightforward statement, usually, 'They must be created by UFOs. What do you think?' There is nothing for it but to reply, 'They could be.' Many circles and rings are connected with UFO sightings, see page 176. UFOs are claimed to be capable of producing the most extraordinary behaviour and phenomena. Their control of force fields unknown to us may well result in rings and circles. It may well be within the capability of a UFO to manipulate a rotary force field which is enclosed in a sharp cut-off electro-magnetic shield. It is also possible that UFOs are only visible when they wish to be in our light spectrum, so the forces they may control could be demonstrated with or without their presence.

Some of my theories have appeared in the worldwide magazine *Flying Saucer Review*. Through this a reader sent me an article by Tom Johanson from the *Spiritualist Gazette*. It expands the parallel worlds theory which I have long supported.

Alfred North Whitehead wrote that science was blind because it dealt with only part of the evidence provided by human experience. There was much more to the world of atomic matter. Arthur Young, a brilliant American physicist, agreed with Whitehead. He said that the 'whole man' would never be fully understood by studying man's physical shape alone. Werner Heisenberg, a German physicist, wrote about the atomic (material) world that atoms are much more than things. He thought there was something else which proved that there is no clear distinction between matter and energy. Another well-known German physicist, Max Planck, spent many years investigating the sub-atomic realm and concluded that another world of reality existed beyond our present world of sense.

Undoubtedly, one of the most startling discoveries was made by two quantum physicists, Professor Segre of Italy and Dr Chamberlain, an American scientist, who discovered the anti-proton. The present sensory world is constructed of protons and electrons which form the atom. Discovery of the anti-proton meant that the proton exists in two forms: protons with mass (touchable matter), and anti-protons (anti-matter). Therefore the world we now see consists of the normal atomic matter, but within that world exist sub-atomic particles: in other words, a non-mass world interpenetrating a world of mass.

Arthur Young went further in his investigations into the sub-atomic world. He concentrated on blending quantum (sub-atomic) physics with the phenomenon of consciousness, and on the strange, little understood phenomenon of light. Light is without mass. Light is like colour. Colour is without mass. For instance, you cannot imagine colour without a vehicle, such as flowers, paint, coloured water, dye, and so on. Therefore colour is without mass, yet it is there. Light is also without mass and without charge or energy, yet is capable of creating protons and electrons with mass and with charge. Young then discovered that light had photons. Photons are a pulse of light which have no mass and no charge, but are able to create protons with mass and charge, proving that two different worlds exist together, one mass and the other anti-mass.

The greatest of all physicists, Albert Einstein, proved that photons exist, but they do not exist in time. In other words, they do not recognise time. He also stated that everything in existence is based upon the photon. The photon has an infinite lifetime, and can take any form it wishes as mass and as anti-mass. With the anti-proton, or photon, which has no mass and no charge, an infinite lifetime is established as the vital bridge between the state of being (now) and the state of 'not-being' in the physical world.

This would seem to support the theory that the circles are created by an unknown force field manipulated by an unknown intelligence.

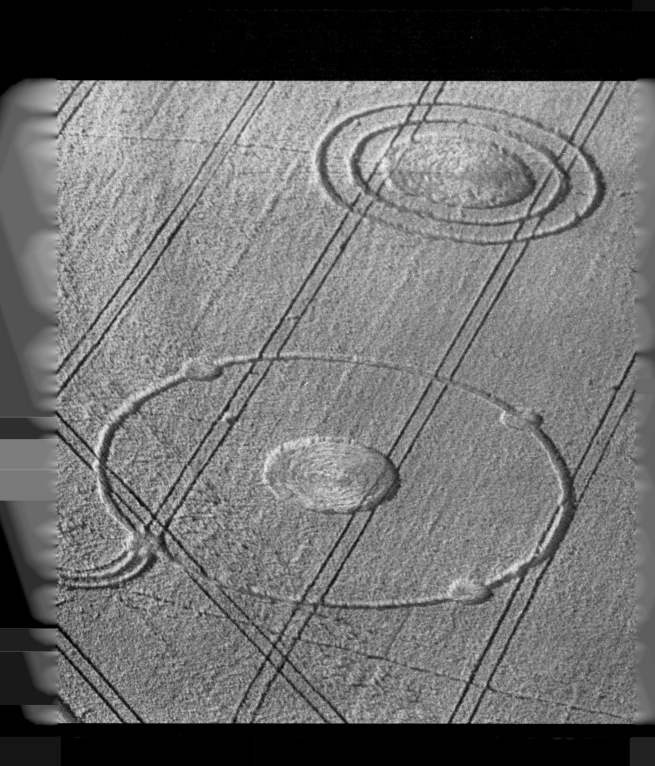

CHAPTER SIX

FURTHER OBSERVATIONS

MYSTERIOUS EVENTS

To give credence to the idea that unknown powers or forces are connected with this subject, startling incidents have happened on several occasions.

While Colin and I were measuring the smaller of a pair of circles at Bratton on 28–9 July 1987, the compass we were using was on the ground in the centre of the circle. We had completed the radial measurements for north and north-west and were about to take the west measurement when we noticed the compass was spinning in a counter-clockwise direction. We could hardly believe our eyes. Then it suddenly stopped and it never happened again. We had not touched or knocked it to make it spin, but we wondered if a steel tape we were using might have influenced it. So, although we always hold the tape end about 40 to 50 centimetres above the compass to avoid any influence, we put the end of the tape on the compass and tried to spin it. Complete failure. We had measured dozens of circles in this way and this was the first time this phenomenon had occurred. What had spun the compass? Something unknown to us had manifested itself.

During a second visit I made to the circle in Dog-leg field, near Winchester, at about 7 pm on 4 July 1987, another strange incident occurred. I was half-way between the centre and the west wall, looking at the swirled floor and the few single standing stems. It was a calm, sunny evening with a light, cool breeze from the north-west. I was facing south when I was suddenly startled by a loud hissing, crackling sound, which gave me the impression it was jumping over the tops of the crops about 10 metres away on my right, covering a distance of about 8 metres. The noise was similar to a high-voltage discharge but far more prolonged and with a pulsating beat. It did not stop suddenly, but faded. I was nonplussed.

Because of this, I took a tape recorder to the field the following evening to see if it would happen again. It was much later in the evening, about 10.30 pm and almost dark. I placed the recorder on the ground near the west wall of the circle and recorded for forty-five minutes. I neither heard nor saw anything, except light traffic noise in the distance from about six cars which passed during that time. I re-ran the tape when I arrived home and for about thirty minutes listened to normal background hiss and slight hum, interspersed with the faint noise of the passing cars. Then a peculiar roaring sound of fairly low frequency started quietly, increased in volume and faded. This happened three times. The third time was loudest, and then it faded away altogether. This noise, or anything similar, is nowhere else on the tape, so I repeated the performance again two days later. This time there was nothing. I suppose I will never know what made this noise, but it was very eerie.

The phenomenon of a microphone picking up sounds that were not audible at the time of recording is not new. Many instances have been described, but that does not detract from the importance of each occasion.

On another occasion we were investigating a circle at the Town Farm site at Westbury on 8 August 1987. Colin was photographing the western quarter of the circle while I was stooping to examine the plant roots along the northern wall. Suddenly, there was a loud knock, as though someone had tapped the lid of a wooden box with a hammer. My head was slightly above the standing crop and the sound seemed very close, about 6 metres away just above the crop. I thought somebody was there who had made this sound. I asked Colin immediately, 'Did you hear that? What was it?' He had and was just as startled as I was. We discussed what we had heard and agreed about the location and description of the sound. We will probably never know what caused it.

During a visit, on 3 August 1987, to the group of three circles beneath the White Horse at Westbury, another strange recording incident occurred. The circle configuration was a large elliptical area with two small satellites on the northern side. The lay was mostly counter-clockwise, but rough with a number of standing tufts. Colin placed his recorder with remote microphone in the centre of the ellipse and, while we measured and photographed, recordings were made on both sides of a 90-minute tape. When this tape was played back in November Colin noticed a peculiar knocking had been recorded. Again, it was similar to a wooden box being tapped and occurred at irregular intervals. The knocks sounded singly or as pairs with a one to one-and-a-half second space between them. The intervals varied from quarter of a minute to three minutes and the singles and doubles were irregularly mixed too. The loudness varied, even with the doubles. Thirty-two doubles and eleven singles were recorded in forty-five minutes and they gradually faded in volume during the last third of the tape.

Being closely associated with these noises creates wariness when you are actually in the circles. It emphasises the presence of unknown forces that must be all around us, manifesting their presence in unconventional ways.

During our visit to the circle in Dog-leg field, Longwood Farm, two independent incidents occurred simultaneously. Colin, Busty, my daughter, Jan, and I were busy measuring, photographing and studying details. I had taken several photos and was at the end of the film. I walked over to the side of the ring and about a metre into one of the tramlines to replace it. I knelt down over my accessory bag. After completing what is a simple task with the Nikon FG, I wound the film on a couple of times, intending to bring it to the first frame position. Then the winder jammed. All I could do was open the back to see what had happened. I was horrified to find that although the film was unharmed the shutter curtains were *buckled upwards* and the camera was useless.

At the same time, I could hear loud words coming from Busty. His video camera had jammed for no apparent reason. It would not run whatever he tried with the controls. After a few minutes, it operated perfectly normally again and, since then, has never malfunctioned.

The camera repairers were unable to understand how my camera's shutter curtains could have buckled upwards, that is, towards the back of the camera. It is easy stupidly to push the shutter down with your finger, but I had not touched them. Anyway I had wound the film on two frames with no trouble.

It seems certain that something significantly unusual occurred in and around that circle for a few moments. Whatever it was cost me £90 and after that incident I made sure I was well away from the circle while changing the film.

You will have read in Chapter One about other isolated incidents occurring during our circle investigations. Blue flashes were seen at Goodworth Clatford in 1985 and Westbury in 1987. Classic UFO shapes seen on film taken at Westbury in 1987 and fork-like shapes at Chilcomb in 1987 only add to the mystery. Colin's experiences at home after the Childrey 1986 report may or may not be connected, but no force so active has recurred since then and, we hope, will not happen again.

UFO CONNECTION?

Well if these are intelligences, then they know something about the physical world that we don't know, and they also know something about the psychic world that we don't know — and they're using it all.

DR ALLEN J. HYNEK

We are always asked whether circles and rings are the result of UFOs. We all have guarded views about this, with varying degrees of agreement. However, there have been reports of UFOs in the vicinity of circles, see pages 35, 68 and 115. There was evidence of possible UFO connection with a circle at Findon in West Sussex, where a swathe was cut through the tree tops, leaving twigs and small branches littered on the ground below. An open mind is healthy, especially when associating with the unknown, when all kinds of possibilities arise.

Here is the first account of an unusual incident that occurred during the summer of 1987. I received a telephone call on 30 August from a woman in Dorset. Her daughter had sent her a cutting of an article by me about the Westbury circles. She told me that after reading it she recalled something she had seen at the beginning of the month.

First, though, she told me about herself. Both she and her husband were retired and she was well into her eighties. She had been a keen bird watcher and had a very good eye for detail. 'You have to notice detail quickly with bird watching,' she said, and, 'I always keep my binoculars handy, even now.' Her bungalow has a lounge window facing south with a lovely view of the sky.

'It was soon after sunset,' she went on. 'I was sitting at my window looking at the beautiful purple-blue sky, when I noticed a very bright golden star. I thought it was a star at first until I referenced it against a group of stars. Then I could see it was moving towards me. I picked up my binoculars and through them I could see it was not a star, it was something round with very bright golden lights. The lights were not around the edge but were inward from it. At the point of six and seven on the clock the lights were blocked out by something projecting downwards. At the point of one and two on the clock there was a projection outwards with small, bright, golden lights on it. At the end of this projection was something like a ball or globe and it had a lot of small, bright, golden lights on it. It was all very brilliant.

'I rushed out into the garden to get a better view,' she continued, 'as it had come right overhead. It was then I called to my husband to come and see. In my excitement I had forgotten that he was now blind and I also forgot my binoculars. I rushed back in to get them and when I came out again the bright, golden object was still there. I watched it again through my binoculars and as it passed overhead and receded, the bright projection was now on the left-hand side. I looked around for somebody to share the sight with me, but there was no one. When I looked up again the object had disappeared, so I decided to follow the route it had taken among the stars with my binoculars, and sure enough it came into view again. This time the object seemed to be in skeletal form and the golden lights were only half as bright as before. The little lights on the globe were like a lot of hundred watt lamps. A thought seemed to be put in my head that it is coming down. It really is coming down and it will leave marks on the ground. Now it began to grow fainter until it disappeared while still seeming to be coming down. I know it was coming down somewhere over Salisbury Plain, well north of here.

'While it was travelling overhead I seemed to be impressed with the thought, they really are going to be different this year, it is different this time. What something was trying to tell me, I don't know, at least I didn't know until I read about the circles, and then I knew that was it. I had a great urge to go to Westbury to see the marks on the ground. I didn't know where Westbury was, I have never been there. I asked my daughter to take me in her car, which she did a few days ago. We hunted around most of the day trying to find the circles but without success. Finally, we asked a farmer where they were and he advised us to go up on the escarpment above the White Horse and look down into the fields. We did this but I was very disappointed because the light was fading and,

although my daughter saw them, I'm afraid I couldn't and I wanted to so much.

'I really wanted to see the marks made by the UFO I had seen.' That was how she ended her story.

This woman is intelligent as can be seen by the technically correct way she carried out her observations and by the manner in which she described them. It is a pity the episode had a fruitless ending for her.

SCIENTIFIC INTEREST

Members of the various branches of science are all interested in this subject, and we are frequently asked to supply information about circle details, especially photographs. However, we do not readily part with our often hard and expensively gained knowledge. We have been approached by analytical chemists, physicists, astronomers, geologists, nuclear research engineers, psychoanalysts, general practitioners, meteorologists, and probably many others who have not divulged their profession while talking to us at various sites.

A teacher wrote to me from Coronach in Saskatchewan, 'Although I am not aware of any of these circles in our wheat fields or pastures here along the United States border, I am a high school science teacher and would appreciate it if you could send me some general information and perhaps even a picture that I could use in the classroom.' On this occasion we obliged and sent her enough information for her pupils to start a project.

Professor Archie Roy of Glasgow University was enthusiastic about the subject when he first met me, an enthusiasm he later projected during a radio interview. A few months after that, when asked for his views about circles by the *Observer*, he is alleged to have declared they were all hoaxes. I fully believe he thinks otherwise, but perhaps he does not want to be involved with a persistent enigma.

This is a transcript of Radio Solent's interview with Professor Roy.

Radio Solent: It seems a full scientific evaluation is to be carried out on the Hampshire rings. It is to be masterminded by Professor Archie Roy from the Department of Physics and Astronomy at Glasgow University. He's on the phone now. Professor Roy, how did you first hear about our rings?

Professor Roy: I first heard about them in an article by Pat Delgado and the photos and plans struck me as being of great interest and so, because I was down in the south here (the south of England), I took the opportunity of going along and having a long discussion with Mr Delgado.

Radio Solent: So what were your initial feelings about them?

Professor Roy: Seeing the plans and hearing the history of these rings and also having a look at two of them in a field, well three of them actually, it seemed to me that while it was not impossible that they could be made by man, the fact that there had been a demonstration that you could make such a ring does not necessarily mean that they are all made that way. In fact, one could say that the fact that there are counterfeit coins in circulation doesn't mean that all coins are counterfeit.

Radio Solent: What form would your investigation take if it went ahead?

Professor Roy: I think that in all scientific investigations which are starting out from some seemingly unknown phenomenon, you simply have to enter the stamp collecting phase. In other words, you collect as many varieties as possible, record them, get all the information you can and then you begin to classify them. After that, if you're lucky, you'll hit on some sort of test, some sort of theory that you can apply and, if it is a good theory, it has predictance and you should be able to predict, in certain circumstances, what the phenomenon should be. At the present state, you see, we haven't the faintest idea, but I must say that some of the explanations that have been put forward, you know, like whirlwinds or viruses or down-draughts from helicopters, they seem to me to be totally inadequate with respect to the beautiful symmetry of these rings and the regularity.

Radio Solent: Now why is it necessary to know what makes the Hampshire rings appear anyway, because farmers take it all in their stride, don't they?

Professor Roy: Yes, indeed. One hundred and fifty years ago, people took it in their stride that if you rubbed cat fur against amber, or things like that, you'd get sparks. People following that up, like Faraday, Clark and Maxwell, have given us the modern world with its technology.

Radio Solent: Now, if you find out what causes the rings, can it have any implications for us, do you think?

Professor Roy: It is impossible to say. All actual phenomena have implications, and in this particular case, I don't see any reason to doubt that it might have far-reaching implications. The trouble is, you see, at the beginning of any scientific investigation, you really don't know where it's going to lead. Lord Rutherford, when he was splitting atoms at the Cavendish, said, 'Of course, this will never be of any practical value whatsoever.'

Professor Roy's words echo all the open-minded thoughts of thinking people on this subject. He does not allow conventional science to bias his thinking. Consequently, as a highly respected professional in his area, his approach to our subject is refreshing. There is just one remark of his I do not agree with. Referring to the circles, he said, 'It seemed to me, that . . . it was not impossible that they could be made by man,' and so on. I am sure he would not have said this if he had carried out the same thorough investigation that we have. He would have discovered all the very complicated characteristics that, we suggest, defy duplication by human beings.

Even so, although he was not fully acquainted with it, he realised this phenomenon has deep implications and justifies thorough, prolonged scientific investigation. During a visit to my house in summer 1986, we were deep in a discussion of the intriguing details of circles, when he made a very encouraging statement. He said he thought the subject justified a professional, scientific investigation and that he would consider approaching colleagues in the Royal Society. This move would surely make possible the provision of suitable equipment for surveillance, both at night and in the day.

Colin was also present and we were both excited by the thought of what future research could produce.

DOWSING PENDULUM REACTION

Dowsing came to mind because of the strange noises and incidents that had occurred up to the end of July 1987. If residues of whatever force had created the circles were still active, it might be possible to detect them with dowsing methods.

I had practised some dowsing in the past, but without success. I had used forked twigs, wires and free-rotating horizontal wire rods, but the results usually left me in doubt as to whether I had detected something or my nerves were simply twitching. When it came to trying this art in a circle I used the horizontal rotating rods, but they swung to left and right and did not appear to make much sense. Then I thought about the pendulum method and made one from an M12 (½ inch) steel nut and a length of strong thread. The results were immediate and startling.

I first tried it out in the Punch Bowl's group of five circles on 25 July 1987. I stood half-way between the centre and the northern wall of the central circle. The bob weight was hanging on about 0·5 metres of thread. For a few seconds it hung quite motionless, then it began to rotate in a clockwise direction. It gradually increased its described circle diameter until it settled at about 40 centimetres. I stopped it and it performed in just the same manner. I repeated this several times with the same result. I even started it off in the opposite rotation, but this quickly died and then it commenced its initial rotation. This happened in four different parts of the circle. Then I tried it in one of the 3·5-metre-diameter satellite circles, where it would only rotate in a counter-clockwise direction, the same direction as the swirl. Two of the satellites had a stronger effect than the others.

To make sure I was able to observe and judge the diameter of the circle described by the bob weight, I had tied a knot in the thread about 0·5 metres from the bob end. Some circles caused the pendulum to swing in a wide, lively way if I shortened the thread by several centimetres, just as though I was tuning it to suit the conditions.

One exceptional case occurred at Westbury, the site of the large ellipse and two satellites. In the ellipse the pendulum swung with a mixture of counter-clockwise rotation and a backwards and forwards swing. I was puzzled until I realised the lay was composed of half each of a swirl and almost radial, and contained many standing clumps, which is probably why the pendulum appeared agitated.

After this peculiar experience in the ellipse, I walked over to the satellite on the north-eastern side and stood in the centre with the thread length at about 1 metre. The bob did not move. It felt like a stiff vertical rod and, as such, forced me to carry it in the direction of the ellipse. When my arm approached the wall of the satellite, the rod-like thread suddenly turned at right angles and moved in a southerly direction. It would have carried on in this direction and taken me right out of the small circle, so I grabbed the thread with my other hand and ended the exercise. I repeated this twice more with the same results. The pendulum did not want to swing at all. It remained rod-like each time, moving sideways and pulling me with it.

I now find the pendulum can give me a 'feel' for a circle. Not only does it always comply with the swirl rotation and patterns, but the energy displayed seems to fit the general surrounding atmosphere.

During the summer of 1987, I extended my experiments with the pendulum. My wife, Norah, daughter, Jan, and I decided to include the Rollright stone circle in Oxfordshire in a mini tour we planned. It is always exciting to walk into this 32-metre-diameter circle with its 70 or so stones of differing shapes and heights. There were about a dozen other people there including 4 or 5 artists who were sitting at various points and painting the view of the circle.

I started operating the pendulum at the first stone I came to. The stones are all numbered on a map and I think it was number 14 or 15. With the length of the thread at about 30 centimetres, the pendulum swung clockwise at a lively rate. The next stone to the left caused it to rotate counter-clockwise. This alternating sequence continued for about 10 stones. The stone, about 1·5 metres high, I was now operating on 'felt' different. The pendulum was going clockwise in a very strong way. The thread was almost horizontal when my forearm was forced backwards and forwards in rhythm with the pendulum, with some considerable force. I could only stop it by turning completely around.

I have included this story because the force that seemed to take control of my arm had exactly the same feeling as the force that caused the thread to 'stiffen' in the Westbury circle. That, too, took control of my arm.

WORLDWIDE REPORTS

The United States

1 On 12 July 1969, in Van Horn, Iowa, a mysterious circular patch of shrivelled soybean crop was discovered 131 metres in diameter. It was not quite a perfect circle. On the north-western side there was a slight elongation, similar to the pear-drop shape described on page 120. Intense heat or radiation was attributed to the affected crop.[1]

2 There are many cases of burned rings in grass areas. A classic example occurred in 1970 in Connecticut. Returning home one night from a lecture, a student saw a large circular craft descending on a field in front of her. She said the craft had flashing red lights and windows. She is unsure how long she stood watching it.

However, when she returned the following morning to the area where the landing had taken place, she discovered a large perfect circle of burned grass.[2]

3 During September 1984 an incident occurred at the Foster Tree Farm in Wisconsin. While walking through a plantation of pines, Mr George Kind came upon a circular area of bent, almost flattened trees 29 metres in diameter. The actual number of effected trees was difficult to ascertain due to the way the swirled condition had piled them up, but was about 75. About a quarter of the trees were broken and had died. The severely bent trees were still growing and their lengths were up to 14 metres. The case was investigated by professionals, but no satisfactory answer as to what caused the phenomenon has ever been put forward.

Circle sites worldwide

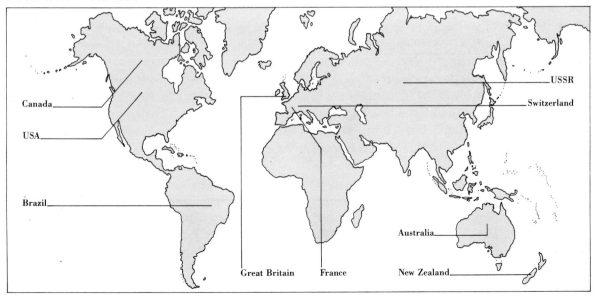

Australia

1 The mention of circles in Australia immediately brings to mind the town of Tully, North Queensland. On 19 January 1966 tractor driver, George Pedley, who was employed by the farmer, Albert Pennisi, was approaching a landlocked stretch of water known as Horse Shoe Lagoon at Euramo, near Tully. He was startled by an object rising from the swampy lagoon, which was hidden at that time by trees. After a shallow dive, the object made off at the classic high speed. Arriving at the lagoon, he was amazed to see an area of flattened swamp grass. He noticed the grass was flattened and swirled in a clockwise direction, and its diameter was about 10 metres. Later that afternoon Mr Pedley told Mr Pennisi and a friend about his find. They returned to the lagoon and verified his findings.

News of this UFO 'nest', as they are known in Australia, soon spread and many investigators visited the area. Four more circles were found in the lagoon, and in each one the plant roots had been pulled out, complete with soil. The plants were floating in the swirled direction. At least one circle was swirled counter-clockwise, the others were clockwise. The diameters of the circles ranged from 2·7 to 4 metres. The extracted and swirled plants were scorched in some cases and observers maintained that the plants discoloured much more quickly than they would have in natural courses of extraction.[3]

2 In the same lagoon on 20 February 1987 Mr Pennisi discovered another five circles. These were all swirled clockwise and much of the vegetation within the markings had blackened and appeared to be dehydrated. A sixth circle was discovered closer to the farm buildings in a waterway.[4]

3 At Tooligie Hill, South Australia, in December 1971 a 2·7 metres in diameter, 10 centimetres deep, circular trough was discovered in a wheat field. The trough was bare soil and the wheat inside was flattened with a counter-clockwise swirl. On the previous night, a large ball of red light had been seen by a local farmer, descending on the field where the trough ring was found.[5]

4 At Bordertown, South Australia, in 1973 seven circles were discovered in an oat field. Their sizes ranged from 2·3 metres to 4·8 metres in diameter and they were all swirled counter-clockwise.[6]

5 At Wokurna, South Australia, in December 1973 one circle was found in a wheat field. The crop was flattened in a counter-clockwise swirl. There were also bare soil patches in the circle.[7]

6 At Port Neill, South Australia, in March 1977 a trough ring of bare soil was found in a field of thin grass, 4·6 metres in diameter. After much investigation no explanation was found for any of these phenomena.[8]

7 Four miles south of Karawinna, near Mildura, Victoria, four witnesses, who were returning home just before midnight on 25 September 1976, saw a group of red lights apparently resting on the ground. The lights began flashing, rose up, hovered and, after drifting away, disappeared after a series of flashes. Some of the witnesses returned to the spot three days later and found a 23-metre-diameter circle of discoloured grass. A bush at one side of the circle was scorched on the side facing the circle. The farmer who owns the land stated that one of his cattle was missing on the morning after the incident. One had also disappeared the previous week.

The circles and rings mentioned here are mostly connected with UFO sightings. Evidence of landings is particularly plentiful. Few aspects of the UFO are as tantalising as the scorched or flattened prints it has left in the soil, crops, snow and ice, probably in every country on earth. Rings and nests have been photographed and analysed hundreds of times. The burn marks, the whirlpools of flattened grass and corn crops, all suggest that some incongruous craft has landed, or at least someone or something wants to let us know that such an event has occurred.

8 A particularly intriguing pattern of rings, whose appearance was preceded by an aerial overture, was reported from Mount Garnet, south of Mareeba, Queensland, in February 1977. There were five perfect circles in the dice-dot pattern, with a centre circle 6 metres in diameter. The other four, symmetrically square and equidistant from it, were 1 metre in diameter. They were located in a hollow, the site being invisible from the road. A near-by tree was found to have been knocked over.

Near-by trees were damaged at Findon, West Sussex, where a set of five circles appeared in 1985. Residents of Mount Garnet said dogs had been behaving strangely for days

and television and radio reception had been shredded by static.

9 Ground marking activity took place in a powerful way at Leitchville, Victoria from late 1977 to March 1978. A meticulous investigator, Paul Norman, visited the farmland of Tom Church at Leitchville on behalf of the Victorian UFO Research Society. He was shown a ring 8 metres in circumference with a burned rim 35 centimetres wide. The physicist and technician who advises the society analysed the soil from the circle and reported that neither fungi nor radioactivity were present. The ring had been created by an electrical burn. A bull had bent an iron bar it was tied to in adjacent farm buildings. On 4 December 1977 Mr Church discovered that near-by neighbours had found circles on their land as well. A 12-metre-diameter circle was found on Lloyd Naylor's property, as was another on Gil Pickering's farm. Mr Pickering has taken a number of colour photos of the circles on his land.

10 More circles were photographed on a farm track 10 kilometres from Leitchville. Paul Norman reported they were precisely the same diameter and equally spaced 30 metres apart. The circles followed the contours of the bumpy road and never varied from their 12 millimetre depth.

11 Over the Christmas period in 1974 an eerie aerial light haunted the small farming town of Narrogin, Western Australia. During this period a flattened oval was found in a wheat crop on the property of Mr F. Chadwick, about 12 kilometres east of Narrogin. The oval measured 18 metres on the major axis and 12 metres on the minor axis.

12 In 1977 in Orange, New South Wales, scores of reports were made describing the aerobatics of a strange light in the sky during one day. The following day a burned and flattened triangle was discovered on a farm. A physicist visited the site and conducted some tests. His compass went berserk (remember Bratton, 1987, page 00). The triangle had been highly magnetised, and the grass was found to be water resistant.

13 A UFO landed on the property of farmer Bill Boulton of Mildura, Victoria. It was seen by Mrs White who also saw it take off. Later the area was searched by several farmers, who found a perfect 9-metre-diameter flattened swirled circle. The grass was discoloured.

14 During August 1974 at Jandowae, north-west of Toowoomba, Queensland, farmer George Nauschutz found a circular depression 2 metres in diameter and 30 centimetres deep. This depression contained five holes bored into the ground and these holes changed direction underground. They also contained a white powdery substance.

15 Two farmers, George and Viv Huckle, discovered a circle in a patch of sacaline on their property at Forbes, New South Wales, while ploughing. It was located near a waterhole, 45 metres from a solitary tree and less than 1 kilometre from the farm buildings. They were astonished at the way the plants had been destroyed. At the centre the earth was bared. Towards the perimeter the thistles were swept down in a counter-clockwise direction. Some of the stems were pulped.

16 In October 1976 a farmer in Torrita, Victoria found two circles about 7 metres apart, each 5·5 metres in diameter, on his property. Colour photos of these circles have been studied by researchers who concluded that they seemed to have been created by considerable force. The grass was burned and swirled. Beyond the well-defined edges of the circle, the grass was neither burned nor swirled. Four days before this occurrence, a horticulturist, Mr Jones, had reported finding a 9-metre-diameter circle of singed grape vines. The vines bore no fruit that season.

17 On a farm owned by Mr W. Errat of Boggabri, New South Wales, a hole scorched in the earth was discovered. It was agreed by investigators that it was similar to others photographed in New South Wales and Victoria. This hole was 1·8 metres in diameter and about 15 centimetres deep, with a large 45-centimetre-deep hole in the centre. Several smaller holes were found around the edge of the main depression. This circle contained a fine white powdery substance like the Jandowae, Queensland, circle.

18 In February 1976 at Kettering in Tasmania, a good observation of a UFO landing on rough grass alongside a sports field took place. The witness watched the craft on the ground for about 5 minutes from a distance of about 25 metres. It then rose and streaked away at high speed in the usual classic manner. The witness returned to the landing spot the following day and noticed a circular area of the grass had been scorched. This circle was about 7·5 metres in diameter.

Numerous other rings and circles have been found in other parts of Australia over many years.

Brazil

In Ibiuna during the early hours of 17 June 1969, observers saw what looked 'like a brilliantly illuminated window'. This window, hovering above the ground, seemed to be curved and was approximately 9 metres long and 3 metres high. The light illuminated the ground for a limited distance and a spotlight picked out some trees and shrubs. It was observed for about 45 minutes, remaining stationary, and then vanished. The ground beneath where the light had been was examined during daylight and a circle of flattened grass, 8 metres in diameter and swirled counterclockwise was found. The floor area also contained some small secondary swirls.[9]

Canada

1 In 1974 at Langenburg, Saskatchewan, a farmer came across a small disc-shaped UFO spinning in a clockwise direction in a field of rape. As he got closer, he saw there was not one but five of these objects sitting on the rape. They then lifted off simultaneously upwards to a height of about 60 metres, where they hovered over him for a while and then shot off at speed. On the ground he found five circles of clockwise spiralled plants.

2 In 1977 at Odessa, Saskatchewan, a 5-metre-diameter circle was discovered in a field of grass. The grass inside the circle was completely charred. There was some gradation around the edges and a 1-metre-diameter vein of charred grass in the circle's central area.[10]

During 1986 and 1987 I was asked by the Canadian Broadcasting Corporation to talk about circles and rings on a programme called 'Quirks and Quarks', which was transmitted throughout Canada. The following are extracts from letters I received as a result of the programmes.

From Sidney, British Columbia. 'My experience happened about 6 years ago, in a meadow being grown for hay, I think about 10 acres. It was in the middle of May as the grass was nearly 2 feet high as far as I can remember. We hay here earlier than in England. I was walking along the headland with my dog, everything was silent about me, blue sky and sun, when this almighty bang came and my dog jumped, wondering what on earth was happening. It was a much louder bang than a gun. I have belonged to shooting clubs for years so know what they sound like. The grass was going around in a clockwise direction, banging and cracking, and the grass as wild as could be. I thought it would be torn out by the roots. We stood there and watched. Only 8 metres from where we were, with a blink of an eye shall we say, there was dead silence, all as quiet as before, no sign of wind or any movement of trees. I walked over to the circle and stood looking at it, the grass was limp and partly flattened, not torn a bit, it just looked tired. The circle was as far as I can remember about 3 metres around with upstanding grass all around like the rest of the meadow, just that one circle that was now different from the rest of meadow. I did not notice any spiral pattern in the centre, it all looked the same. After studying it for a while, shook my head and made some comment to my dog about finishing our walk around the meadow. No sign of disturbance in the rest of my walk or anywhere else in the meadow.'

From Hugh Cochrane, Toronto, in his book *Gateway to Oblivion*. 'I came across a number of reports of these rings, most of which occurred in low land near the lake Scugog area of Ontario, about 50 miles north-east of Toronto. This was back in the early 1970s. Several reports in 1978 were of flattened circles in tobacco plants around Delhi, about 60 miles south-west of Toronto. Since the above reports involved swamp grass and tobacco plants it would appear that the phenomenon is not limited to grain crops.'

From Belleville, Ontario. 'Have you come across rings in corn maize, not wheat, before? The rings or other shapes have been here before, but the cause may not be piezoelectric since the corn may be too strong for the effect. Corn is the main local crop.'

From Ottawa, Ontario. 'With regard to the mysterious rings on the ground in England, Claire Strutt of Eganville, Ontario, took aerial photographs (coloured slides) of perfect circles apparently burned into the grass at the time of flying-saucer sighting reports in that area in about 1971–2. He showed those slides at an Experimental Aircraft Association, Chapter 245 (Ottawa) meeting.'

From Sarnia, Ontario. 'I have examined three circular impressions, two in a pasture/woodland environment in the Pontiac County area of Quebec and one near Grand Bend, Ontario. The Grand Bend effect was located in a corn field. The Pontiac impressions are located about 9·5 and 20 kilometres from Shawville, Quebec. This is just across the Ottawa River from Renfrew County, Ontario. Let me begin by describing one Pontiac impression, which is nearly identical to the second Pontiac impression. Examination of the site revealed a donut-shaped impression of dead grass on a green, living background. The outer ring was about 9 metres in diameter while the inner hole of the donut was 7 metres across. Within this 1-metre-wide ring, the dead grass was swept in a clockwise bend. It appeared that the grass died from severe induction heating of the soil. The stems were not damaged, but the root structure of the grass was destroyed. Grass within the hole of the donut was green and healthy. The ring-like impression was located on a 5° slope below a large ironwood tree. Whatever produced the ring also burnt the branches of the tree. It appeared as if a "vehicle" of some sort descended from above, forcing down the branches of the tree, burning them and trapping them between the vehicle and the ground. The soil and grass were not as severely damaged in the area below the trapped branches. Soil sampled from inside and outside the ring was normal. At the inner edge of the ring (bordering the donut hole), two granite rocks showed temperature discoloration, as if scorched by induction heating. Two young ironwood trees, (about 80 centimetres in height) located inside the outside edge of the ring, were bent and deformed and contained severe burn marks, but they were living and growing. The farmer explained that two or three similar-sized poplar seedlings had died. Poplars have shallower root structures than ironwoods and apparently suffered the same fate as the grass. This impression was at least two years old when I first saw it and the effect is still visible, about ten years later! The individual who first discovered the donut impression said that it appeared overnight. He knew this because he had crossed this section of pasture as a short cut to reach his carpool each morning and had returned late each afternoon by the same route. I am sure that no one reaped financial gain from this oddity, so a financially perpetuated hoax is out. The farmer reported that 'government men' took samples at the site but no conclusions were published. The above incident takes on an eerie twist because UFOs were reported in this area at the time when the impressions appeared. Quite a number of local residents reported visible sightings and certain sightings were witnessed by more than one observer at the same time.

The second donut was found near Ladysmith, Quebec. Ladysmith is 20 kilometres from Shawville and about 16 kilometres from the first site. This impression is also on a 5° slope next to a large ironwood tree and in a secluded location. The dimensions of the donut are the same as the first one, or possibly a bit smaller. Apparently this incident occurred in late December, in the early evening, with snow on the ground. A villager, about 1 kilometre away, phoned the farmer who owned the property where this second impression was found, informing him that there was a fire in his bush. Apparently snow, steam, reflected, refracted and scattered light simulated the effects of fire. According to the report, the barn blocked the farmer's line of vision and he observed no flames or smoke, so he pursued the matter no further. About half an hour later neighbours reported a brightly-lit stationary aerial object over near-by Gray's Lake. After five minutes or so, the object accelerated away at high speed into the south-east sky.'

From Lac du Bonnet, Manitoba. 'In my last parish in Langenburg, Saskatchewan, we had quite an experience with these rings. One of my parishioners was harvesting canola [oil seed rape] on a September morning when, in his own words, he came across about five UFOs in his field, rotating, and then they took off and disappeared. Needless to say this event sparked a great deal of media hype and he underwent a great deal of personal ridicule and abuse.

'However, the rings were solid evidence and were exactly as described on your show. As well, I believe they also appeared on a neighbouring farm.

'I disagree that these were caused by turbulence, however, as no such weather conditions were in evidence. What I do believe might be a cause of these rings is piezoelectric forces that could have affected my parishioner, causing him to hallucinate these UFOs. The swirling nature of the phenomenon his brain registered suggests that some sort of vortex force was involved which generated an image quite like what he imagined a UFO would be. At any rate the fact that several cattle and dogs were also greatly disturbed during this "flap" suggests that some force was acting on them that perhaps is similar to what animals sense prior to earthquakes.

'I am not a physicist but I did want to share this letter with you. The date of this was in the late 1970s. Newspaper accounts and photos are available from local newspapers and the local television station also took film of the circles.'

From Calgary, Alberta, an unusual incident with an unknown force. 'I was driving down a gravel road near Arrowwood, Alberta, when suddenly about 100 feet in front of me and off to the right in a fallow field a great explosion of dust took place. The soil went into the air at about 30 feet and fell back to earth with slight drifting in the direction of the wind. I stopped the car and went back to examine the area. My first thought was that a seismic crew was working but that did not prove to be the case. It was not a whirlwind or dust devil since the dust fell gently back down to the ground. Upon close examination of the area where this took place, I could find no markings whatsoever on the surface of the field. My perception was that this dust explosion was no more than 1 metre in diameter. After examining the area and finding no markings, I truly began to feel that I had been seeing things.'

From Fauquier, British Columbia. 'I saw such a phenomenon in 1980 in an expanse of marsh grass near Coeur d'Alene, Idaho. About 6 metres in diameter with very sharply defined edges, no feathering. Soil was a rich muck over an arm of Coeur d'Alene Lake which had been filled by erosion and stream deposits. Almost a

muskeg. If your correspondent in Britain is interested I have a photograph which I can search out from my files for his use. There were also several other smaller circles in the same area, but I will need to find the photos to refresh my memory as to size and shape.'

From Labrador City, Newfoundland. 'Our permanent home is now at New Ross, Nova Scotia, a country place but without any wheat-lands – the crop is Christmas trees. However, there is much meadowland and behind our house a large meadow where I believe there was once a stone circle. On that meadow strange lights have been seen at night – not marsh gas as it is a flattened hilltop and very dry. Some of the lights seemed to take the form of what are called UFOs. They seemed to be attached to circular objects that rose from or landed on the ground. Some of them would rise and stay in one place in the sky all night, or would move about in various directions including from west to east, or would appear to be playing leapfrog with each other. I examined the ground where one had appeared to land but the grass was too short for me to be certain that any mark or depression had been left. However, another man, a farmer, who lives about 100 miles from us did find a circular mark on grass on his property. So UFOs are one possible explanation.

'With regard to UFOs, these are not always visible, especially in daylight. I am pretty certain of this because when one passed over during the night, it would often affect the time on our clock – in fact we were all 'turned back' in time, even in our conversation or the radio, TV or stereo. This sometimes also happened in the daytime, but although we looked we could see nothing in the sky even on a clear day. The UFOs we experienced and sometimes saw were elliptical in shape, and it occurred to me that they might have been made of the thought-energy I had also been seeing, though I can't say for certain that I actually saw one forming. The nearest I came to that was when one suddenly appeared on the meadow, initially it was a ring of light.'

From Pembroke, Ontario. 'My observations of some mysterious rings were made in 1967, in Grattan, County Renfrew, Ontario. On a flat area, adjacent to the shore of Garvin Lake, I had noticed these conspicuous rings which were almost perfectly circular (almost too precise to be caused by nature). The site, an old abandoned farm property, was grown over by species of wild grasses, within which two very distinct rings were visible, side by side.

'These two rings were approximately 4 metres in diameter and perhaps 1 metre wide. The grass which had grown within the ring area appeared to be totally dried up and sort of stunted or tramped down. There was evidence that the site had been used by deer hunters, who had erected makeshift tents. My thoughts were that these rings were caused by hunting dogs, which had been tied to a picket in the ground and then ran around in a circle, tramping all the remaining grass.

'Several months later, a farmer near Chapeau, Quebec discovered similar rings of about 8 metres diameter in his pasture. The grass there appeared to be burnt, according to his report to the *Pembroke Observer*. I never visited the site. However, this incident led me to look more closely at the rings I had discovered, but I could not find evidence of either burn marks nor any indication that dogs were ever tied up there. As I often returned to that site for hunting and fishing purposes, I could notice that the grass appeared stunted for another two years. In the third year these rings could no longer be distinguished.'

England

1 On the Yorkshire Moors in 1978 two soldiers of the Royal Armoured Corps saw a UFO while driving a Land-Rover. It was hovering about 50 metres away when their engine stopped. After a period of about five minutes the UFO left and they were able to start the engine. The following morning they returned with a Sergeant and discovered a large circle of burnt grass where the UFO had hovered.[11]

2 At Elgin, north-east Scotland, two young girls saw a UFO hovering in a wood. They described it as glowing with a red light and there was a silver-suited man standing by it. The mother of one of the girls said she remembered hearing a strange whirring sound at about the same time as the girls' sighting. When the girls took their mother and a neighbour back to the landing site they found a circular patch of flattened grass.[12]

3 In July 1963 a number of people were astonished to see a 2·8-metre-diameter circular depression in the centre of which was a 1-metre-deep hole. It had appeared overnight in a potato field at Manor Farm, Charlton, Wiltshire. Radiating from its centre were four slits. The circle was bare earth only.

4 During the famous Warminster UFO flap in the mid-1960s, when hundreds of sightings were reported, the following story emerged. Along the road between Sutton Common, Norton Bavant and Heytesbury, Wiltshire, in wooded areas usually near marshland, a number of 'landing spots' were sighted by author, Arthur Shuttlewood. In every case, reeds and grass had been curiously flattened in what invariably seemed to be clockwise fashion, blades swept smoothly inert in shallow depressions. It was significant that most circles, depressed and clearly formed, measured exactly 9 metres in diameter, and some more than this.[14]

France

A report from St Souplet (Nord) states that a woman and her son heard a noise outside the house late one night. Peering out through the unshuttered window the woman saw a red light hovering over the spinach patch. She called her son to the window and they watched the light until it soared away. In the morning they found a 3-metre-diameter circle with the plants pressed to the ground and undamaged.

New Zealand

On 4 September 1969 near the town of Hamilton, New Zealand, a 13-metre-diameter circle was found in scrubland. The growth was dehydrated and radioactive. Three deep depressions, 3 metres apart and forming a triangle, were found in the central area of the circle. All the vegetation surrounding the circle was healthy. Horticultural consultant, John Stuart-Menzies said, 'Some object appears to have landed on that spot and taken off again, emitting some kind of short-wave high-frequency radiation. This seems to have cooked the plants instantly from the inside outwards. I know of no earthbound source of energy capable of creating a circle in this manner.' Nuclear scientists called in by the government said the cause was root rot and blight.[15]

USSR

Three scientists were camping at a site about 100 kilometres from Moscow in the late 1970s. One night, when in their tents, they heard a loud babble of voices. None of them recognised the language and they all felt a sense of unaccountable fear. It was half an hour before they dared look outside and there stood a shining, violet-coloured object about 25 metres high, looking like a giant electric light bulb. It rose, swayed slightly, then soared upwards into a fluorescent cloud. Next morning, the campers found a circle of flattened grass about 180 metres from their tent where the UFO had stood or hovered.[16]

Switzerland

In the fascinating book *Light Years* by Gary Kinder,[17] the main character, Eduard 'Billy' Meier, discovers a number of swirled circles in grass. These circles or 'tracks' are claimed by Meier to be created by UFOs with which he has apparently had dozens of close encounters of the fourth kind, when he has been abducted.

The description of the circles in Kinder's book echo and underline the descriptions contained in this book. For instance, Meier is showing two investigators two three-circle sets: 'We walked almost 100 metres through the wood and came to a place where the wood was closed all around and there was a meadow in it. There was very high grass, maybe little less than a metre. There were six landing tracks from the two ships. The grass in the circles was turned counter-clockwise. The interesting thing was the grass was not broken. If you break grass it will lie down, but this grass was not broken. Untouched, the rest of the grass in the meadow still stood stiff and tall all around the 2-metre-diameter counter-clockwise swirls so precisely pressed down.'

The descriptions of the circles emphasise how amazed they were that these flattened swirls could be pressed into the tall grass without crushing it and that the grass never wanted to rise up again. It continued to grow in the horizontal position. Further descriptions explain how it was thought the grass plants were induced and were now growing with a horizontal magnetic orientation rather than vertically as in a normal field.

CONCLUSIONS

The hard facts about circles and rings are all there, openly displayed and waiting to be understood. Each site, wherever it occurs in the world, provides similar or fresh evidence to that already encountered. Each new formation strengthens the current feeling of many people that we are dealing with something which hints at some form of manipulated force. Nothing in the current state of conventional science can account for all that has been described in this book.

The circles are given the same treatment as UFOs by many governments – they either openly debunk them or they show no interest.

I do not know what I may appear to the world, but to myself I seem to have been only like a boy playing on the sea-shore, and diverting myself in now and then finding a smoother pebble or a prettier shell than ordinary, whilst the great ocean of truth lay all undiscovered before me.

SIR ISAAC NEWTON, 1642 – 1727

In theory circles could be considered to violate our ground space as UFOs violate our air space. But, unlike UFOs, the circles are not a fleeting phenomenon which can be misinterpreted or ignored. Many are on view for weeks, even months. When it is finally established what creates these ground markings, even though our ground space has been theoretically violated, no real harm has been done to us. Crops will have been lost in some cases, but the circles and rings are creations of beauty, and credit should probably be given to them for uniting people in the study of a common interest.

If an event occurs which causes substance or material to be displaced, then science, and physics in particular, is involved. When it occurs regularly to produce circles and rings in crops, then high-level scientific investigation should be carried out and in such a way that interested parties, the world over, are kept informed of events and progress.

Our group has carried out a small amount of scientific investigation which has been limited by our personal funds. We are now in a position to advise on what we consider necessary for pre-circle formation, formation itself and lengthy post-formation programmes. Even at our limited rate of exploration, fresh details are constantly forthcoming, a situation which will no doubt continue.

It is our hope that the scientific establishment at large will realise that the puzzles encountered in the circle phenomenon are already part of the very fabric of science, so that serious research can begin.

Conclusive evidence and final answers are a long way off. However, while completing this book we have pursued one extremely interesting line of research which may well be the way forward. We are continuing with our endeavours in the hope that our patience will be rewarded.

A NOTE ON THE AUTHORS

Pat Delgado is a retired electro-mechanical engineer and Colin Andrews is chief electrical engineer with the Test Valley Borough Council. They are founder members of the Circles Phenomenon Research Group.

Notes

1 Dinah L. Moche, *Life in Space* (Ridge Press, 1979).
2 Ibid.
3 *Australian Flying Saucer Review* (November 1966).
4 Australian Centre for UFO Studies, September 1987.
5 *ACOS Bulletin*, no. 11 (September 1977) – journal of the Australian Co-ordinated Section Centre for UFO Studies.
6 Ibid.
7 Ibid.
8 Ibid.
9 *Flying Saucer Review*, vol. 16, no. 1 (January–February 1970).
10 Moche, *Life in Space*.
11 Nigel Blundell and Roger Boar, *The World's Greatest UFO Mysteries* (Octopus, 1983).
12 Ibid.
13 W. Raymond Drake, *Gods and Spacemen in Greece and Rome* (Sphere, 1976).
14 Arthur Shuttlewood, *The Warminster Mystery* (Neville Spearman, 1967).
15 *Spacelink*, vol. 6, no. 2 (1970).
16 Blundell and Boar, *The World's Greatest UFO Mysteries*.
17 Gary Kinder, *Light Years* (Viking, 1987).